Concrete
A Homeowner's
Illustrated Guide

Concrete
A Homeowner's Illustrated Guide

David H. Jacobs Jr.

TAB BOOKS

Blue Ridge Summit, PA

FIRST EDITION
SECOND PRINTING

© 1992 by **TAB Books**.
TAB Books is a division of McGraw-Hill, Inc.

Library of Congress Cataloging-in-Publication Data

Jacobs, David H.
    Concrete : a homeowner's illustrated guide / by David H. Jacobs,
  Jr. — Trade ed.
      p.   cm.
    Includes index.
    ISBN 0-8306-3913-6
    1. Concrete construction—Amateurs' manuals. I. Title.
TA682.42.J32 1992b
693'.5—dc20   91-38538
                                                CIP

TAB Books offers software for sale. For information and a catalog, please contact
TAB Software Department, Blue Ridge Summit, PA 17294-0850.

Acquisitions Editor: Kimberly Tabor
Book Editor: Lori Flaherty
Director of Production: Katherine G. Brown
Book Design: Jaclyn J. Boone
Cover Photography: Susan Riley, Harrisonburg, VA
Cover Design: Cindy Staub, Hanover, PA   TAB1

Contents

Acknowledgments

Seldom can a concrete job of any size be accomplished without help. The same holds true for writing a book about concrete work. Therefore, I would like to thank a few special people for their assistance and support throughout this project.

Keith Rieber and Gerald Brewster took pictures of some of the concrete jobs featured in this book. I appreciate their time and effort, as it is not always easy to stop in the middle of projects to photograph progress.

Gratitude is also extended to Dave Reum and his crew from Custom Concrete for being so patient while I photographed them during concrete forming, pouring, and finishing operations. The same goes for the Gamby brothers.

The people at Southern Oregon Concrete were equally helpful, spending time with me to explain their operations and how the concrete business is viewed from their perspective.

Thanks to Van and Kim Nordquist from Photographic Designs for doing an excellent job processing the film and printing pictures, taking extra time to make sure each photo turned out just right.

Janna Jacobs provided lots of needed support and plenty of help with the overall organization and photo coordination.

I must also thank Kim Tabor from TAB Books for her support and editorial assistance. Her enthusiasm and professional insight helped to make this an enjoyable project.

Introduction

*I*n the last few years, concrete has become so expensive that some have referred to it as *gray gold*. Because of the cost and professional labor fees, many homeowners are going without concrete walkways and patios. If your concrete job would cost only half as much as a professional's estimate, maybe you could afford to get it done right away.

Most homeowner's concrete needs are about the same—a walkway down the side of a house, a driveway, or a patio in the backyard. These are rather simple jobs that conscientious do-it-yourself homeowners can easily do.

Far too often, homeowners have expressed that they would like to do concrete flat work jobs themselves, but ... Their "buts" are, for the most part, simple problems with easy solutions. Almost 50 percent of the cost of most concrete flat work projects goes toward professional labor fees, and because form lumber and stakes can be used for future projects to offset their initial expense, do-it-yourself concrete finishers can realistically expect to save up to 50 percent off professionally installed concrete.

Concrete flat work includes projects such as, patios, walkways, driveways, and mowing strips. Books about masonry sometimes include a chapter or two on concrete flat work, but they generally don't go into great detail about flat work designs, concrete yardage calculations, order delivery, pouring, finishing, and other basic concrete flat work.

From page 1, this book takes you through each and every step you'll need to form, pour, and finish quality concrete flat work. From designing jobs to final cleanup, everything is described, explained, and illustrated. You'll learn how to design slabs and walkways that conform with your yard's landscape, how to properly form creative designs, determine how much concrete is needed, how to estimate overall costs, determine the amount of help you'll need, how to pour and finish concrete flat work, how to protect it, and how to add some custom touches.

You might not qualify as a concrete professional after reading this book—you simply can't learn all there is to know about a professional trade by reading one book—but by following directions and thinking jobs through, you'll be able to complete professional and functional concrete flat work.

Should you need help, information is included on what type of help you can expect to find in your area. In many cases, a simple phone call to the right person can get your questions answered and problems solved quickly. If you are not sure about how a particular phase of your concrete project is supposed to be completed, don't be afraid to ask for help or advice. Professional concrete dispatchers are generally quite knowledgeable about their trade and seldom fail to offer useful advice to customers with special problems. If need be, especially for truly unique flat work projects, ask a professional concrete finisher to survey your job site and offer solutions to puzzling questions. Their fee for this service should be quite minimal and well worth it in the long run.

This book was written for you, the do-it-yourself homeowner. Follow the instructions, advice, and recommendations to form, pour, and finish your own concrete work and you're sure to get the self-satisfaction of completing projects on your own, as well as saving money.

Chapter 1

Design considerations

You must allot ample time for planning the type of concrete job ultimately desired. So, get a chair, pencil, paper, and something cold to drink. Then, sit near the proposed work site and start drawing plans (FIG. 1-1). They do not have to be perfect or to scale—just usable and easy to read. Sketch a number of various designs. Add or delete planters, experiment with different widths and lengths, and consider what the overall job will look like with the existing or proposed landscape. Keep in mind that innovative concrete flat work designs can go a long way toward improving the general appearance of almost any landscape while still providing usable paths, entertainment areas, or work spaces.

WALKWAYS

Although basically simple in design, walkways can be attractively designed to offer more than just a suitable means to get from the backyard to the front yard. For example, leaving room for a planter next to your house or a fence would be more attractive than butting it up against the foundation (FIG. 1-2). Leaving open spaces for flowers and shrubs breaks up solid concrete with greenery and color. There are several factors that must be considered first, however.

Will a particular side of the house get enough sunshine to support plants? Will visitors see and use the walkway? Will it simply be a space to store garden tools and trash cans? For the latter, a wide walkway extending from the foundation outward might be best to maximize storage space and still allow for plenty of yard space (FIG. 1-3.).

Width

Intended walkway use helps determine its width (FIG. 1-4). Most walkways are 3 feet wide, ample space for pushing a wheelbarrow or lawn mower

12' × 20' = 240 sq ft
240 sq ft ÷ 80 = 3 yards of concrete

Needed -   3 yards concrete
            3 - 12' 2×4 stringers
            1 - 20' 2×4 stringer
            2 - 4×4 wet post anchors
                (stirrups)
            3 workers - 1 on chute
            2 on screed

Garage

Note: remove section of fence to let truck through

Fence

House

12'
enough room for concrete truck

Sliding door

2" below mudsill high point

12'

12' stringers
20' stringer

Stirrups
3' from side
1' from end

Low point   20'

1-1   Thoroughly plan your concrete job with rough blueprints. Include measurements, concrete yardage calculations, and any special notes.

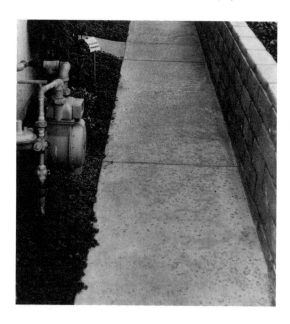

1-2 A planter separating the house and walkway. Note how concrete runs downhill from back to front along the block wall.

I-3 The concrete provides a sheltered walkway under an overhead deck while providing maximum yard space for lawn and other landscaping.

I-4 Most walkways are generally 3 feet wide but can be wider to better integrate with existing conditions such as an entrance corridor or covered breezeway.

across. Four feet or more is uncommon but might be necessary for specific needs or landscape conformations (FIG. 1-5). Wide walkways are actually mini-slabs; quite useful for outdoor furniture when used as a front or back porch (FIG. 1-6). In some cases, you might want to break up the plain design of wide concrete walkways by inserting treated, 2-×-4 wood stringers, bricks at specific intervals, or your own personal design.

Narrow walkways of less than 3 feet wide are primarily for decorative purposes and are only functional for one person to walk on at a time. This type of walk is best suited for infrequently used pathways with little traffic.

Add an extra inch for 3-foot-wide walkways. This extra inch will not make much difference in appearance but will come in handy during pouring and finishing. This is because most large concrete tools—a tamp, bull float, and fresno—are 3 feet wide (FIG. 1-7). An extra inch easily accommodates these wide tools so that they can be used down walkway lengths in a single motion as opposed to across walkway widths just 3 feet at a time.

For example, if you form a 3-×-20-foot walkway, it will be much quicker and easier to push tools down the 20-foot run in one maneuver

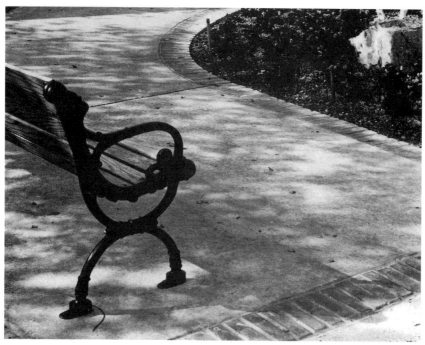

I-5 Besides serving as functional walking surfaces, extra wide walkways provide plenty of space for benches, outdoor furniture, planters, and other accessories.

I-6 This extra-wide walkway entrance doubles as an attractive and dimensionally symmetrical front porch area. Notice how its brick border and inserts help concrete flat work blend with surrounding brickwork.

1-7 Consider forming walkways at 3 feet 1 inch so tools such as the fresno can be operated lengthwise.

than to finish concrete across the three foot width, covering only a 3-foot section each time.

Slope

Water runoff is a very important and essential design element for any concrete project, especially when slabs are located near structures. Ideally, concrete walkways and slabs should run downhill toward streets, drains, or natural runoffs (FIG. 1-8). Proper slope guarantees that water will flow away from certain areas, like next to house walls, and toward specific drainage sites. As you sketch design plans, determine where you will want

1-8 Patio slabs and other projects should slope downhill away from your house and other structures toward drains or other natural water runoff areas.

water runoff to go and incorporate those features into the job. Under no circumstances should you ever want rain or water from any other source to run toward your house or other structure. Always be certain that, concrete slopes slightly away from walls and toward the yard or street.

To help you determine the proper slope, use the mudsill on your house as a guide (FIG. 1-9). This is generally a lip located at the bottom of walls where stucco or siding stops near the top of the foundation. This lip should be very close to level. Use it to determine a high point when forming walks designed to run downhill. The space between the top of the concrete and the mudsill will be less at the high point and much greater at the low point, and water will run toward the low point.

1-9 Concrete was poured using the stucco mudsill as a guide. This slab is level under the far wall's lip but slopes down along the left wall.

Many concrete jobs have had to be broken out and replaced because water runoff was not sufficiently factored into the design. Water puddles next to house walls or in low spots around yards negates the primary reason for having concrete walkways, especially during rainy weather. Besides, pools of standing water next to walls and foundations will eventually lead to serious dry rot, mildew, and other dampness problems, especially when water is left in place for prolonged periods.

Along with planning for normal rainwater runoff at ground level, don't overlook runoff from roofs. Investing in rain gutters can solve some runoff problems if they drain away from enclosed planters and into specific drains (FIG. 1-10). Bordered planter boxes without drains can quickly become saturated, accumulating water to pool and flood. Consider installing three-inch diameter drainpipes under walkways to help divert water runoff from roof gutters into drains or toward the street.

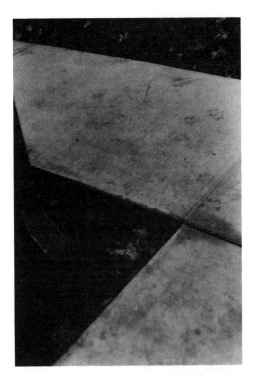

I-10 Water flowing out of this rain gutter downspout would surely flood the planter area had a flexible plastic drainpipe not been installed under the walkway.

To prevent mud splatter against your house as a result of rainwater falling from the roof into planters, place concrete close enough to the house so that water drips on the concrete first and then gently falls away. To be sure that concrete is poured close enough to the wall, spray some water on the roof and measure how far from the house the drips land; then form the inside concrete edge accordingly (FIG. 1-11).

PATIO SLABS

Some of the considerations mentioned for designing walkways are also appropriate for patio slabs, especially slope. It is essential that you ensure that the entire slab slopes away from walls and toward open-yard areas and/or drains.

Slope and slab depth are two factors that go hand in hand. Regular patio slabs do not need to be any deeper than 4 inches. Normally, concrete is poured to the top of 2-×-4 forms. When determining a high point for slope consideration, take a good look at the natural slope presented. It is sometimes easier to pour slabs along the natural grade (FIG. 1-12). If, however, the grade is not proper, you will have to remove dirt from the high side to fill in the low side. Dig out only enough to make adequate room for your 2-×-4 form. Excessive digging is unnecessary and only creates more work.

To help determine grade while minimizing excavation work, use 2-×-4 screed boards as guides. Small wood stakes are nailed to the top edge of a

1-11 Walkways poured away from house walls should be positioned so that water drips falling off the eaves land on concrete rather than splattering into planters.

1-12 This patio slab conforms to the yard's natural grade—a downhill slope running toward the bottom of the photo.

2 × 4 cut to fit inside a formed area. With stakes riding on top of the forms, the body of the 2-×-4 screed will quickly show you when enough dirt has been removed or filled in to make a perfectly even, 4-inch slab depth.

Other features you might consider for patios include open planters, 2-×-4 wood stringer inserts, and means of support for patio roofs (FIG. 1-13). You might even consider pouring two separate slabs of different dimensions connected by a short walkway. For all intents and purposes, any design is good as long as it is functional, channels water away from walls, and meets with your visual approval.

1-13 Small concrete pads were poured outside the slab to accommodate patio cover supports. Installing posts this way provides more open room in the patio area.

MOWING STRIPS

Often, maintaining a well-manicured yard requires some form of separation between grass, flower beds, decorative gravel, mulch, and other landscape features. Lengths of 2-×-4 treated lumber or benderboard work fine for some landscapes, but a more permanent and long-lasting solution is concrete mowing strips.

Mowing strips are generally 4-inch wide concrete curbs poured and finished at ground level or left to stand a few inches above grade depending on their intended function. They can be used to separate gravel driveways from lawns, function as borders around trees, or be placed along fence lines to give lawns a tidy border and make grass edging easier (FIG. 1-14).

Unless there is a special need to partition different ground cover root

1-14 Mowing strips can be rectangularly formed or made to curve around circular driveways and other rounded landscapes. Strips of benderboard are used to form curves.

systems at specific depths, like between Bermuda grass and flower beds, mowing strips do not need to be much deeper than 1 to 2 inches. Mowing strips should be at least 3 to 4 inches thick but you can adjust their tops to be flush with grade level or allow them to stand higher like a curb or raised border. Mowing strips can be formed in rectangular shapes with straight-form lumber or in curved patterns by using thin benderboard forms to match landscaping designs.

Placing concrete inside mowing strip forms is usually done with a shovel. This is because forms are generally not wide enough to allow direct wheelbarrow pours. And, because their dimensions are seldom bigger than 4×4 inches, not much concrete is needed. Small strips may be poured using bags of concrete mixed in a wheelbarrow, while larger projects might be best supplied by 1-yard haulers (described later) or with concrete left over from another job.

EXPANSION JOINTS

Believe it or not, concrete actually expands and contracts with variations of temperature. To help prevent cracks along walkways and slabs, finishers install expansion joints at strategic points (FIG. 1-15).

An expansion joint is created when a piece of special felt, benderboard, or other treated lumber is placed in concrete as a means to absorb the expansion and contraction of concrete as it heats or cools (FIG. 1-16). These inserts are especially important along walkways and at other places where slabs of concrete intersect or butt up to other slabs; like, the edge of a driveway poured up to the edge of a sidewalk.

Expansion joints should be placed about every 10 feet on walkways

I-15 This crack could have been avoided if an expansion joint was installed at the corner just to the right of the post.

I-16 The expansion joint on the right side and the seam running perpendicular to it are correctly installed and should be provided on all walkway corners.

and at every point where corners are formed or intersections occur. When a walkway extends from a patio slab, for example, an expansion joint should be inserted in line with the outer edge of the larger slab. If not, a crack will eventually develop across the walkway at the point where it butts against the larger slab.

Plan ahead for the adequate placement of expansion joints to be sure they are inserted at equal intervals. This ensures that they conform to a uniform pattern. On a long walkway measuring 27 feet, for example, put in a joint at the 9-foot mark and another at the 18-foot mark. Each section will measure 9 feet and look uniform.

Expansion joint felt is available at lumberyards and ready-mix concrete plants (the place where concrete delivery trucks are filled). Although most city building codes require $1/2$-inch felt be placed between sidewalks and driveways and any other city owned concrete slab, you could opt for redwood benderboard or treated 2-×-4 lumber expansion joints on other concrete projects around your house.

THE FINISH

Concrete slabs with a smooth surface, like garage floors, might look nice but are dangerously slick when wet. To prevent slabs from becoming as slippery as ice when wet, finishers enhance traction by applying any one of a number of various concrete finishes. By far the most common are broom finishes, which are very easy to accomplish and hide many finishing errors. Small lines left by hand trowels are quickly wiped away by broom bristles, leaving their own distinctive line pattern. These broom lines not only make slab surfaces look neat and uniform, they add a great deal of positive traction on wet concrete surfaces for pedestrians.

Broom finishes can feature more than just straight line patterns. Brooms that are wriggled as they are pushed and pulled across surfaces create wavy line designs. With a hand-held foxtail broom, you could even design a series of half-circle lines similar to the effects windshield wipers make on auto windshields.

Broom finish texture is determined by how wet concrete is when a broom is applied over its surface. The wetter the concrete, the more defined and deeply textured the lines will be. Conversely, operating a broom over a more cured surface results in very light lines. Heavier-textured concrete surfaces offer maximum traction. Consider heavy lines for steep driveways that experience auto traction problems. For walkways and patio slabs, light broom finishes generally offer plenty of excellent traction for everyday foot traffic, even when concrete is wet.

INTEGRATING DESIGN

Concrete projects can be designed to conform to almost any landscape scheme. If your yard is decorated with lots of shrubs and flowers, you might prefer to leave small areas open inside your patio slab for plants or trellis features. Walkways don't always have to be straight to be functional. Some creative plan sketching might result in a unique blend of offset walkway sections around trees or bushes, adding flair to an otherwise mundane landscape.

Along with the concrete projects featured in this book, a lot of creative ideas are offered through home improvement magazines and other house and garden periodicals. Take a walk through your neighborhood to see what kind of unique concrete designs have been incorporated into landscapes similar to yours. Better yet, you might even have fun driving through exclusive neighborhoods to find truly one-of-a-kind designs that could easily be duplicated or expanded upon.

Because concrete flat work projects are generally considered relatively permanent additions, take the time needed to design a perfect scheme for your landscape. As you work to set forms and grade, don't be afraid to change designs if need be. It is always much better to take a few extra days to plan and form a perfect slab or walkway with exact dimensions than to pour and finish a project with end results that are less than satisfying, either functionally or visually.

Chapter **2**

Concrete costs

Calculating the cost of concrete is not difficult. There are some variables that must be figured in, but once you understand why and how certain fees are charged, you should be able to estimate delivery costs quite closely. After all of the forms have been set, measure the project's total square footage. These figures will enable you to get a firm delivery price, which should only be altered if the concrete truck remained on your job site longer than anticipated.

BASIC COST

Concrete is sold by the cubic yard, broken down into 1/4-yard increments. One yard of concrete generally sells for around $50, depending on the ready-mix concrete company and geographical location. If, for example, you were to order 1 1/4 yards of concrete, the basic cost would be (using a $50 per yard figure) $62.50—$50 for the first full yard plus $12.50 for the 1/4 yard. Chapter 3 explains how to calculate concrete yardage based on square feet.

On top of this basic per-yard cost, there might be extra charges. Short load and overtime fees are common. Actual dollar amounts and time limits vary among companies. Allow for sales tax, too.

Ready-mix concrete companies operate independently and frequently specialize in specific types of work. Some outfits are often set up to supply large-scale builders with hundreds of yards of concrete per week and prefer to shy away from smaller pours. Other companies, generally the smaller ones, are better equipped to deliver smaller concrete loads and are well suited for common do-it-yourself projects. With this in mind, expect larger companies to charge more for smaller deliveries than businesses that commonly deliver to homeowners.

Call a number of ready-mix concrete companies to determine their price structure. A dispatcher will answer your questions and even verify the amount of concrete needed for your job by using the square footage figures you give him. Along with a basic cost, all of the other fees can be openly discussed and calculated. By calling a few companies, you can

3 yards of concrete ordered
Concrete selling for $50 per yard
3 × $50 = $150

Customer orders 3 yards, takes 45 minutes to
unload, and requests a 1-yard cleanup.

Basic charge:	$150.00 for 3 yards
Overtime (15 minutes):	7.50
Cleanup (1 yard):	$100.00
	$257.50
Charges with no cleanup and correct order	
Basic charge	$200.00 for 4 yards
Overtime (15 minutes):	7.50
	$207.50

A savings of $50 with correct ordering

2-1 Concrete companies allow a certain amount of time to unload concrete. Overtime charges are incurred when unloading time exceeds the time allotted. If extra concrete must be delivered after the initial load, you will be charged cleanup fees.

decide which one can offer the best overall price and service for your project.

SHORT LOAD

Most ready-mix delivery trucks can carry 7 to 8 yards of concrete. To compensate for deliveries that are less than a truck's capacity, companies commonly charge what is called short load fees. Like basic costs, fees vary among companies.

Generally, you can expect to pay about an additional $7 per yard for every yard under seven. If 3 yards of concrete are ordered, it will cost an extra $28 above the basic cost. At $50 per yard, a 3-yard delivery would cost $178—$150 for 3 yards delivered plus $28 in short load fees—$7 per yard for every yard under seven (7 yard-capacity minus 3 yards delivered = 4 yards short load).

Short load fees are not hidden and every concrete company is familiar with them. By shopping around, though, you might find one that only calculates short loads based on 5-yard capacities. In other words, you might only have to pay a $7 short load fee for every yard under five instead of seven.

OVERTIME

Concrete companies allow customers a specific amount of time to unload concrete. After that time has elapsed, they charge overtime. As with short load fees, this is common practice used to compensate for the extra time a truck remains at a job site when it could otherwise be delivering concrete somewhere else. Dispatchers rely on concrete finishers to unload trucks within specific time frames so they can schedule trucks for other deliveries throughout the day.

Some companies allow only 4 minutes per yard unloading time while others might allot up to 10 minutes or more per yard. Some charge a flat per hour fee in addition to actual concrete costs. While gathering concrete cost estimates from ready-mix dispatchers, be sure to ask about their unloading time limits and related fee schedules.

Overtime charges are based on minutes. Rates vary but are generally around 50 cents per minute for every minute over the allotted unloading time. Even though 50 cents a minute might seem like a lot, don't rush your job just to escape this extra fee. An additional 10 minutes at 50 cents a minute will only cost $5, and that might be all the time you need to catch up on screeding or other important chores to guarantee a first-class pour.

A lot of times, concrete truck drivers are in control of overtime fees. In this case, do yourself a favor by getting on the driver's best side by offering him, or her, a cup of coffee or cold drink upon arrival. Fully describe the job layout and point out specific problem areas where concrete placement might be difficult or hard to reach. If the driver recognizes that you are properly prepared, with all helpers and equipment ready to go, you stand a fair chance that a few minutes of overtime might get overlooked.

Getting ready-mix concrete out of a truck and into forms will be the single most physically demanding and important phase of your job. Should you experience problems with concrete placement or screeding, or the project at hand just takes longer than anticipated, don't fret. It is much more important to place concrete correctly the first time than to worry about a few extra dollars in overtime charges.

Concrete is heavy and can quickly tire workers if they have to move it from spot to spot after it's poured; especially to correct mistakes made while hurrying to unload just to beat overtime charges. Your chances of avoiding overtime fees are enhanced when your job site is properly prepared, forms secured and braced as needed, and workers thoroughly familiar with the job before the concrete truck arrives.

CLEANUP

A cleanup load is one that is made after the first truck has made its delivery and there was not enough concrete on board to complete the job. This only applies to jobs where the original truck had the capacity to carry the full amount but the customer made an error and simply failed to order enough concrete. For example, if you ordered 3 yards and that was not enough, the second truck coming with a cleanup load is going to cost a lot. In some cases, just one extra cleanup yard of concrete can easily cost $100 or more (FIG. 2-1).

Contractors frequently rely on cleanup deliveries for extra large jobs when lots of footings, stairs, or other irregularly shaped projects are involved and accurate yardage estimates cannot be fine-tuned to within 1/4-yard increments. For these steady customers, who order hundreds of yards of concrete a week, cleanup fees are waived. However, should a miscalculation on your part result in a shortage of concrete, you will have to pay a hefty cleanup charge in order to get a second truck to deliver more concrete.

The logic behind cleanup fees is simple. If one concrete truck could have easily delivered 4 yards initially instead of just the 3 yards ordered, it could have been on site a short while, delivered its load, and then been on its way to another job. Since it has to make a special trip back to the plant and then back to your house for a cleanup, however, the company has to rearrange its delivery schedule, pay for extra driver time, fuel, and other overhead. The extra cost and inconvenience is made up through cleanup fees.

Frequently, a few small privately owned concrete companies try to be very accommodating to homeowners and do-it-yourself concrete finishers. They gladly answer questions and some will even visit job sites to offer advice, take measurements, and calculate concrete yardage. This type of personal service could be available in your area just for the asking.

DRIVERS

A reference to concrete truck drivers was made earlier in regard to overtime fees. In addition to that, it is your responsibility to confer with the driver so he comprehends exactly how you intend to pour your job. After all, he is the one who will maneuver the truck and control concrete unloading speed.

Drivers depend upon concrete delivery for a living, and it stands to reason that they have witnessed a lot of different pours. Chances are, they have learned a few tricks of the trade by working alongside or watching professional finishers in action. Make sure they survey your work site. Not only will they get a clear picture of the job for driving and maneuvering needs, but they might also be able to pass along a few helpful hints, such as whether certain forms need extra bracing or how concrete placement could be made easier.

Thanks to power steering and plenty of experience, veteran drivers can often maneuver their big trucks into unlikely tight spots. By getting closer to forms, drivers can eliminate wheelbarrowing, which makes placing concrete much easier. Confer with the dispatcher to determine how much room is needed for trucks. Then, if at all possible, provide the space.

ONE-YARD HAULERS

The cost of ordering concrete from a ready-mix plant for jobs of 2 yards or less becomes quite expensive because of short load fees. In such cases, you might consider using a concrete trailer, commonly referred to as a one-yard hauler. A number of concrete companies, gravel yards, and home-improvement centers rent one-yard concrete trailers for as little as $65 per yard, calculated in 1/4-yard increments.

These heavy-duty trailers are filled with fresh concrete mixed at the rental facility through mini concrete machines. Concrete is very heavy, you'll need a stout pickup truck or utility vehicle to pull the heavy load. Don't rely on small cars with snap-on bumper hitches to work safely. If need be, borrow a friend's pickup truck equipped with a heavy-duty dock bumper.

There are no extra charges for trailer rentals and most companies allow two hours unloading time with no overtime fees. Concrete can be ordered in 1/4-yard increments, perfect for small jobs like mowing strips.

Traveling distance for one-yard haulers is limited to about five miles. As trailers are operated over roads, normal vibrations cause rock and sand to settle toward the bottom. Prolonged travel results in a mix of extra rocky material at the bottom and soupy cream on top. Because additional on-site mixing is not possible, concrete mix consistencies could be jeopardized.

Trailers are unloaded in one of two ways. Some styles are equipped with a small door at the bottom edge of the rear trailer panel. A jack mechanism at the front allows users to elevate the front of the trailer to tilt it so concrete can flow out of the door as it is opened. Others simply have a trough incorporated along the top edge of the rear panel. As the jack tilts up the front of the trailer, concrete pours over the top of the back panel.

If you must travel more than 5 miles with a one-yard hauler, ask the batch plant operator to use the smallest rock (aggregate) possible; pea gravel is best. Smaller aggregate measuring no more than 3/8 inches in diameter (pea gravel) has a tendency to stay suspended longer than larger material that generally measures 1/2 to 3/4 inches in diameter. This should result in a much better overall mix once you arrive at your project site.

Along with prescribed amounts of sand, gravel, and water, sacks of cement (the powder that holds everything together) are added to mixes to make concrete. Unless otherwise requested, ready-mix plants usually add five sacks of cement to every yard of concrete. One-yard batch plants, on the other hand, frequently use just four sacks of cement per yard. Although it might cost $5 a yard more, you should always request at least a five-sack per yard mix. The extra bag of cement gives concrete more strength and makes it more creamy for better surface finishing.

With pea gravel mixes, most finishers prefer six-sack mixes. They have found that this richer concrete blend helps small aggregate disperse and suspend better as well as making finishing surfaces much more creamy. This is a valuable tip for one-yard haulers, especially when concrete is used for flat work projects that will be clearly visible as part of an overall landscaping plan.

ONE-YARD TRUCKS

To complement concrete trailers, some outlets offer 1-yard, mini concrete delivery trucks. A small concrete drum is attached to the frame of a heavy-duty dual-wheel truck. It constantly rotates through a chain-driven hydraulic assembly. Travel distances are almost unlimited. About the only concern is water evaporation from mixes and actual travel time as they relate to mixes setting up and hardening in the drum.

One-yard trucks cost more than trailers. This is simply because maintenance and fuel costs are higher. However, you can still expect to save money on 1- or 2-yard jobs when compared to the cost of a regular ready-mix truck delivery.

If your project requires 2 yards of concrete, be sure to section the job in such a way that the first load will completely fill in a specific formed area and not be disturbed when the second load is poured later. You can install an expansion joint at the junction where the first yard stops and the second one starts or make plans for a planter or other solid division at that breaking point. This is an important consideration, as you will not want fresh concrete to butt against material that has already begun setting up, a perfect condition for the formation of cracks (cold joints).

Chapter **3**

Calculating yardage and arranging delivery

The actual mathematics involved in determining the amount of concrete needed for a project is not difficult. If you can add, multiply, and divide, you've got it made. Concrete is sold by the cubic yard, but calculations to determine the amount of concrete needed for a flat work job is done using square feet equations. Remember the old formula for figuring out square feet? Length × width = area. In this case, it will be, in feet:

length × width = area (in square feet).

The square feet is then divided by a constant to give you the amount of concrete yardage needed.

CALCULATING A CUBIC YARD

One cubic yard can be described as a cube measuring 3 feet × 3 feet × 3 feet. To convert the yardage into feet, multiply the sides by each other in feet. For example,

Side #1 × side #2 (3×3) = 9 square feet.

Take this product and multiply it by side #3 (FIG. 3-1.):

9 square feet × 3 feet = 27 cubic feet.

One cubic yard (a yard), therefore, equals 27 cubic feet. One yard of concrete will cover 27 square feet at a depth of 1 foot (FIG. 3-2).

Because your slab will only be 4-inches thick, as opposed to a full 1 foot, you will have to divide those 27 cubes into smaller ones that measure 4-inches thick. To do this, multiply those 27 cubes × 3; because 4 inches is one-third of a foot (12 inches). Those 27 cubes will now cover three times the area when spread out at a depth of only 4 inches (FIG. 3-3). You can now see that a cubic yard will cover 27 square feet 3 times, which

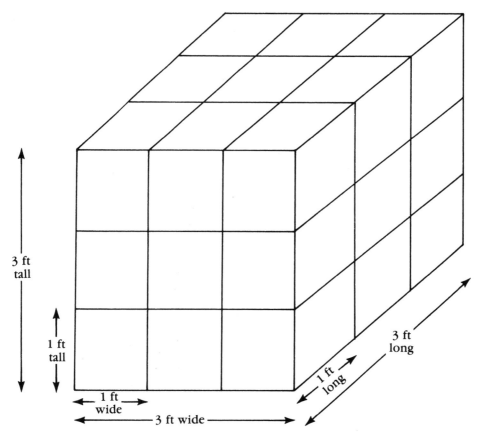

3-1 Equally divided, 1 cubic yard could be separated into 27 separate cubic-foot blocks; 1 cubic yard is equal to 27 cubic feet.

equals 81 square feet ($27 \times 3 = 81$). One yard of concrete will cover 81 square feet at a depth of 4 inches.

There are no variations from this basic formula unless the depth of your concrete project is more than 4 inches. So, don't try to squeeze any more concrete out of a cubic yard. If the slab you have planned measures 85 square feet and the depth is exactly 4 inches, one yard will not quite cover it all (FIG. 3-4). In that case, you should order 1¹/₄ yards of concrete.

ONE-FOURTH YARD INCREMENTS

Although concrete is calculated by the cubic yard, concrete companies are able to sell it at ¹/₄-yard increments. You can order any amount desired, as long as it is totaled in ¹/₄-yard segments. For example, if you need extra concrete to cover a slab area of 85 square feet at 4 inches deep, you will have to order 1¹/₄ yards. There will be a little concrete left over, perfect for a small mowing strip, trash can pad, or step landing. Just be sure the extra projects are formed and ready for "mud" at delivery time.

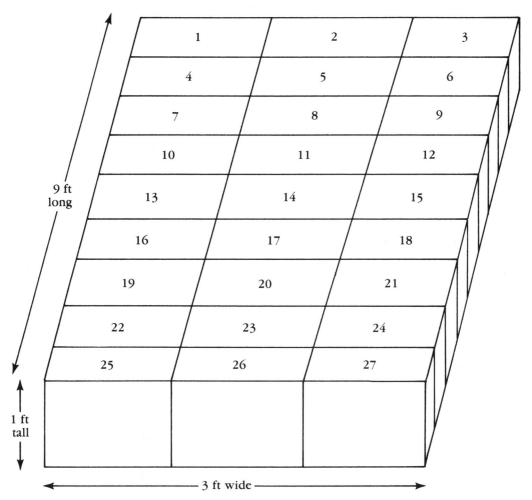

3-2 One cubic-yard of concrete will cover 27 square feet at an even depth of 1 foot.

This same 1/4-yard increment formula holds true for slabs that measure less than 1 yard but more than 3/4 yards. You still have to order the next highest 1/4-yard increment to be certain you will have enough concrete for the job. With that in mind, try to measure out jobs that will most closely engage areas with figures close to 1/4-yard intervals. Tip: 20 square feet at a depth of 4 inches equals 1/4 yard.

Cubic yard shortcut

The number 81, as in 81 square feet per cubic yard at 4 inches in depth, is a difficult number to use as a divider. It will not evenly fit into even numbers, i.e., 200 divided by 81 will not produce an even-numbered answer. You might have a lot better luck using the number 80 instead of 81 (FIG. 3-5).

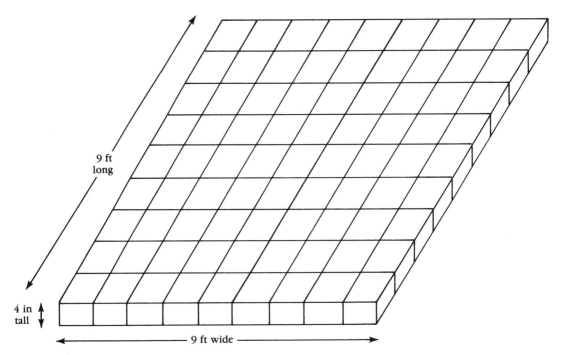

3-3   Concrete flat work is normally 4 inches deep. One cubic-yard of concrete will cover 81 square feet at an even depth of 4 inches.

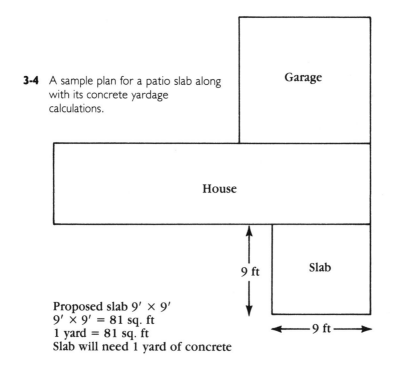

3-4   A sample plan for a patio slab along with its concrete yardage calculations.

Proposed slab 9′ × 9′
9′ × 9′ = 81 sq. ft
1 yard = 81 sq. ft
Slab will need 1 yard of concrete

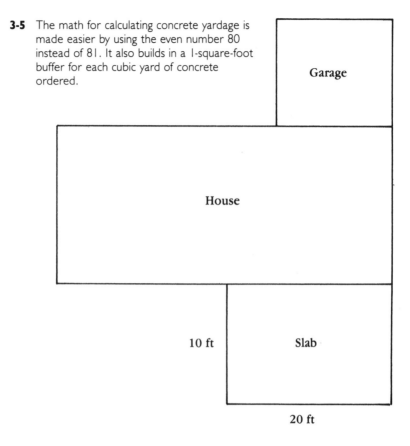

3-5 The math for calculating concrete yardage is made easier by using the even number 80 instead of 81. It also builds in a 1-square-foot buffer for each cubic yard of concrete ordered.

Proposed slab 10′ × 20′ - - - 10′ × 20′ = 200 sq. ft.
200 sq. ft ÷ 81 = 2.469 yards of concrete
200 sq. ft ÷ 80 = 2.5 yards of concrete

Using 80 as opposed to 81 has another benefit besides easier mathematics. It gives you a 1 square foot buffer per yard of concrete. By omitting that one square foot, you have gained a little concrete. This very small amount could be a lifesaver on a job that might not be graded to a perfect 4-inch depth or if a helper should accidentally dump a wheelbarrow of concrete on a job you had calculated perfectly to the 1/4 yard.

Whenever calculating yardage for a concrete job, multiply the length times the width and then divide by 80. Use accurate measurements and overestimate rather than underestimate. For example, if a side of a slab measures 9 feet 10 inches, use a figure of 10 feet even. It is much better to have too much concrete than not enough. The same philosophy holds true for actual yardage. If your figures come out to 190 square feet, round it off to 200 square feet. You'll have to graduate up to the next 1/4-yard increment anyway, so why not make your calculations easier while creating a slight buffer of extra concrete at the same time.

Yardage example

Let's do an example for a slab that measures 10×20 feet. To determine the amount of concrete needed, multiply the length times the width, then divide that answer by 80:

10 feet × 20 feet = 200 square feet

This is the total square feet included in the slab. Remember, it must be graded and formed at a normal depth of 4 inches. Now, because this area is 4-inches deep, we divide the square feet by 80:

200 sq. ft. ÷ 80 = 2^1/$_2$ or 2^1/$_2$ yards of concrete.

Using this formula ensures that enough concrete is delivered to complete a 10-×-20 slab. Again, the depth cannot exceed 4 inches at any point. This cannot be overemphasized. The vast majority of novice concrete finishers who run short on their concrete loads do so because grading is not accurate. Although their math might have been perfect, irregular depths of 5 inches or more simply used up more concrete than anticipated.

One of the major concerns most novice do-it-yourself concrete finishers have is their lack of confidence in ordering enough concrete to avoid any cleanup charges. If you follow these directions and accurately grade to a 4-inch depth, however, you should not have any problems. Also, be aware that most 2-×-4 milled lumber actually measures 1^1/$_2$× 3^1/$_2$ inches, not a full 2×4 inches. Therefore, if you use a normal 2-×-4 board as a guide for grading, you'll really be placing that grade at a depth of between 3^1/$_2$ inches and 4 inches. Also, slabs of 3^1/$_2$ inches fair just as well as full 4-inch-deep slabs for normal use, even for driveways.

AGGREGATE

The aggregate (rock) used to mix concrete is available in different sizes and most concrete companies stock aggregate according to diameter size. Although some companies use aggregate as large as an inch in diameter, most concrete deliveries contain material mixed with ^{3}/$_4$-inch diameter and smaller. Anything larger than ^{3}/$_4$ inch can pose problems for slabs and walkways that require a smooth, good-looking broom finish. Larger aggregate batches fair better for footings and foundations where material is vibrated with a special tool to provide smooth side finishes. Other jobs, like wall grouting and those that use a concrete line pumper, require pea-sized gravel aggregate.

Professional concrete finishers sometimes special order smooth rock as opposed to crushed rock. Smooth riverbed rock is most common on exposed aggregate finishes where the top layer of cement cream is washed away after the slab has been properly finished and set up strong enough to allow finishers to walk on it without making any marks. After the slab is washed off three or four times and the water is clear, smooth rock surfaces are exposed to reveal a truly custom finish.

If your job is planned as a special project, make sure you discuss your aggregate needs with the concrete dispatcher. He or she will be able to address your needs and recommend the right size aggregate for both visual acceptance and strength requirements.

MIX

Although concrete is sold by the yard, it is actually mixed by weight increments. Specific weight measurements of sand, aggregate, and water are mixed with five to six sacks of powdered cement to make a cubic yard of concrete. Cement is a pure powder, while concrete is the term used to describe the entire finished mixture of sand, aggregate, water, and cement.

Ready-mix batch plant operators use a predetermined chart to gauge materials as they are added together to form mixes of concrete. Sand and aggregate in increments of hundreds of pounds are added to gallons of water and then thoroughly mixed with appropriate amounts of cement.

Concrete batch plant operators can alter mixes as needed. Some jobs might require specific aggregate sizes, others might need heavier sand concentrations. Still, some city regulations might require all concrete be mixed with six sacks of cement per yard. This is common in areas that experience severe winters and other extreme climates (FIG. 3-6).

3-6 On the left is smooth rock aggregate measuring 3/4 inch and smaller. The handful on the right is pea gravel. You can specify either size aggregate when ordering your concrete.

Six-sack cement mixes are also recommended for jobs where concrete line pumps are used to pump concrete from the delivery truck to inaccessible areas. These small trailer-mounted pumps use a small-diameter hose and pea-sized gravel aggregate because larger aggregate will not fit through the small hose. The extra rich and creamy mixture of six-sack per yard concrete makes finishing much easier. A sack of cement generally costs about $5. Expect to see that added charge on bills for deliveries of six-sack mixes.

PLANNING A POUR

The overall type of concrete pour you plan will determine the concrete mix and the amount of help you'll need. Helpers are an integral part of any job. Professionals can handle small pours all by themselves because of their experience and expertise. Novices should plan on having some help available, however, no matter how small the slab or walkway.

If you need to use wheelbarrows to get concrete from a truck to the forms, have a number of helpers on site. For a 2- to 3-yard job, it is recommended you have at least four, preferably five, workers. You'll need two people to run wheelbarrows, two to screed and one to help move concrete on the ground to assist the screed effort. If your job requires more than three yards, plan on having one or two more helpers; at least for the amount of time it takes to unload concrete.

Regular, five-sack, 3/4-aggregate concrete is fine for almost any walkway, patio, or driveway. It can be wheeled with no problem and it finishes just fine. Other mixes needed for unique design characteristics, or because of local regulations, must be special ordered.

Should your forms be located in such a way that a concrete delivery truck can dump directly into them, you'll still need a few helpers. One to work the truck chute, two to screed, and another to assist the screed work by moving concrete around as needed with a heavy-duty rake or square-point shovel.

When pouring directly from delivery trucks, you must account for the driver's ability to maneuver. Drivers commonly sit in the cab and move the vehicle forward as directed by the chute man while operating the controls that start and stop the flow of concrete down the chute. Most trucks are 7- to 8-yard rigs. Although they are big, you will be amazed at how some veteran drivers can maneuver them into very tight places.

If your job site appears big enough to accommodate such a vehicle but you are not quite sure, call the ready-mix dispatcher and ask about the dimensions needed for minimal truck maneuverability. It may be well worth the effort to remove a section of fence or temporarily relocate a storage shed in order to get a concrete truck close enough to a job site and eliminate the need to wheelbarrow concrete, especially if your job requires more than 2 to 3 yards.

Larger concrete trucks are also available. These rigs can haul up to 10 yards of concrete at one time. The added weight of 3 yards of concrete requires them to have an extra set of wheels. These wheels are hydraulically raised and lowered. They are placed on the ground while the truck is on the road with more than 7 yards. When it arrives at a job site, the wheels are raised out of the way. Even with the wheels raised, this particular truck is very long. Make sure there will be enough room for rigs of this type to operate around your job site.

Determining when to pour

The decision as to which day to pour your concrete job depends on a few factors. Initially, you must determine when all forming chores will be complete. You must then consider the availability of helpers and the concrete supply company's delivery schedule. Coordinating and confirming these factors is critical. If the forms are not ready, you'll either have to postpone the job or pay standby charges while the concrete truck idly waits for you to fix them. In addition, failure to have enough helpers available could make your pour a painstakingly difficult operation.

Concrete delivery companies regularly schedule jobs as they are called in. Certain contractors frequently have standard delivery times scheduled weeks in advance. Around these mainstay contractor customers, concrete companies fill in schedules with deliveries to other customers. Call the concrete dispatcher with your order a minimum of two days before you want delivery. Optimally, you should schedule delivery one or two weeks ahead of time to be certain the date and time you desire is open and available. By confirming a date and time for concrete delivery this far in advance, you have the best chance of securing the date and time most suitable for you, as opposed to being scheduled in at the concrete company's most convenient time.

For the most part, weekday deliveries are best. This is when concrete plants are operating with full crews. Many companies make Saturday deliveries but commonly charge additional fees to cover driver and dispatcher overtime wages. Fees can range from $5 per yard to flat rates of around $25. Weekend delivery charges are not standard and vary widely from company to company.

Determining the time to pour

Concrete hardens as water and moisture leave the mix. This occurs through moisture absorption into the ground and through surface evaporation. During hot weather, water can quickly seep into parched dirt (the main reason formed areas are always sufficiently wet down before pours) and evaporate at a tremendously rapid pace. When this happens, slab surfaces dry too fast, which can result in cracks or missed opportunities for workers to effect attractive surface finishes.

Conversely, cool winter temperatures combined with seasonally soggy soils cause concrete mixes to retain water and moisture for extended periods of time. Under these conditions, water has no place to go. The ground might already be saturated from rain and cooler temperatures will hinder normal surface evaporation.

Most professional concrete finishers prefer to pour concrete as early in the morning as possible. On long, hot summer days, pouring in the morning allows concrete to harden at a slower rate during cool morning temperatures as opposed to quickly setting up during the heat of the day. During winter, concrete takes much longer to set up and you'll need as much daylight as possible to effect final finishing operations. Pours between 7:00 AM and noon generally work best. Some concrete companies, at a customer's request, will use hot water in the mix to speed curing times.

Generally, figure that concrete will take about four to five hours to cure enough to support final finishing maneuvers during mild temperatures between 65 and 75 degrees Fahrenheit. Cooler temperatures may require more, while hot days exceeding 90 degrees Fahrenheit could cause concrete to flash in less than an hour. During winter, many concrete workers have had to set up temporary lighting to see well enough to finish slow drying concrete late at night. While during hot summers, they have had to work at feverish paces just to be able to get a halfway decent finish on fast curing concrete surfaces. In both situations, early morning concrete pours provide the best opportunities for optimum concrete curing conditions and surface finishing efforts.

Inclement weather

Concrete cannot be adequately finished in rain or snow. Raindrops and snowflakes disturb finishes by adding water to already wet surfaces and diluting the creamy top texture. Unless you have a means of covering slabs with a tentlike structure for protection against the threat of rain or snow, cancel the delivery and reschedule. Concrete dispatchers do not like cancellations but are accustomed to it during bad weather periods.

Professional concrete finishers have been forced to cancel and reschedule concrete pours as many as five and six times due to inclement weather conditions. Although this causes scheduling nightmares, it is far better to be safe than to risk losing a slab finish to the pockmarks, holes, and uneven finish left behind by raindrops and water runoff. Just remember, once concrete is on the ground, you have to stay with it until the surface is finished and hard enough to withstand raindrops without being adversely affected.

Never be afraid to cancel a concrete delivery. Dispatchers are used to these problems and will reschedule at your request; just don't wait until the last minute to cancel. Sometimes, especially when a lot of cancellations have occurred, trucks will run ahead of schedule and could show up at sites an hour early. Once a truck arrives at your job site, you will be expected to unload it, or, at the least, pay a delivery charge even though no concrete was poured. Therefore, determine as early as possible what the weather will be, and call the concrete dispatcher immediately if you decide to cancel.

ORDERING

The first thing a dispatcher will want to know about your job is what type of mix is needed—3/4, five-sack, pea gravel, six-sack, smooth rock, etc. You'll then be asked how many yards are required. If you are not quite sure that your yardage calculations are entirely accurate, give the dispatcher the dimensions of your job and compare both yardage results. If both figures match, you are in business.

Next comes actual scheduling. Once you request a particular date and time, the dispatcher will check a master schedule to determine how your order fits in with other orders on that day. Enough time must be allotted for driving time both to and from your job site, as well as actual unloading time for the amount of concrete ordered. You may be asked to push your delivery time ahead or behind in order to best fit within the master schedule. This is why calling a week or two early works best—because you may be first on the list instead of just a late order used to fill in an otherwise full calendar.

Inform the dispatcher of any special conditions pertinent to your job. Is the site address easy to find or is it off a dirt road out in the country? Will there be any unusual obstacles for the driver to maneuver around? Do you need concrete delivered on the dry side because you are pouring a steep driveway? Factors like these are important and the more you explain your pour to the dispatcher the better your chances are of receiving a timely delivery with excellent service.

Predelivery call

On the morning of your scheduled concrete pour, plan to call the concrete dispatcher a couple of hours before delivery to confirm your order. This helps both you and the dispatcher. By confirming your order, you let the dispatcher know your job site is prepared and the pour should operate in a timely manner. This information assists in overall scheduling and adds an element of on-time service for customers. For you, this call confirms that the address, mix, yardage, and scheduled delivery time are all as expected. You might find that previously delivered orders are ahead of schedule and yours can be brought out earlier if desired or that complications have developed and deliveries are running late.

Chapter **4**

Grading and forming tools and materials

*P*ouring a concrete slab or walkway consists of two basic phases, grading and forming then pouring and finishing. This chapter focuses on the tools and materials required to achieve the first phases.

GRADING

The term *grading* refers to work directed to leveling the ground (base) for a slab to be poured on top of it. A flat base offers the most uniform support for slabs and deters cracks. Flat and even base preparations also ensures accurate dimensions for concrete yardage calculations. Most of the tools needed for grading and forming operations are common household items that most do-it-yourselfers already have on hand.

Heavy-duty shovels are a must. Both round and square points have specific uses, although the majority of work is best accomplished with a square-point shovel. Its straight edge is handy for squaring graded side sections and scraping bases down to grade. A round point shovel is best for digging out trenches and other semi-deep appendages for drains, electrical lines, water pipes, and other items.

A heavy-duty garden rake can be quite useful for smoothing out small mounds of dirt inside formed areas to effect a final grade. It can also be used to gather rocks and other debris strewn about otherwise-even bases. This type of rake also works well in moving wet concrete during screeding.

Unless the ground you are going to grade is easily broken loose and moved around with just a shovel, you will need a sturdy pick. A pick with a point on one end and a wide blade on the other works best. The point works great for breaking up hardpan material while the blade makes quick work of chipping away high spots and cutting out side sections. Hoes are also handy for removing loose dirt or chopping out grassy sections.

The most important and greatly used tool in concrete construction is a heavy-duty contractor's wheelbarrow with an air-filled tire. This is the best and most versatile wheelbarrow for concrete construction. Lightweight garden wheelbarrows with solid rubber wheels cannot stand up to the heavy burden of concrete hauling. If you do not have a heavy-duty wheelbarrow, plan to purchase one at a lumberyard, hardware store, or rent one from a rental company. If your grading job does not require a lot of dirt removal, you can certainly get by without one, but you should have one available for the actual concrete pour; up to three if your job calls for concrete to be wheelbarrowed from the delivery truck to the forms.

Grading can be either a backbreaking chore or a rather simple operation depending on your concrete design, existing landscape grade, and base material characteristics. For the most part, it is easiest to form jobs before grading them. Not only does this clearly distinguish the job perimeters, it also sets grade requirements at the same time, giving you an accurate means to gauge grade depth and dimension progress. Chapters 6 and 7 contain more in-depth information on grading techniques.

FORMING TOOLS

Setting concrete forms consists mainly of carpentry work—sawing, nailing, bracing, and supporting board lumber—and requires an assortment of carpentry tools such as a claw hammer, sledgehammer, tape measure, level, handsaw (power circular saw optional), square, pencil, chalk line, string, nail bag, and other tools.

Claw hammers and sledgehammers

A claw hammer is used to drive nails through support stakes and into forms, as well as for other nailing chores. Although you might be able to get by driving stakes into the ground with a claw hammer, a three-pound mini-sledgehammer works much faster and more accurately. It also helps to prevent wood stakes from splitting after repeated hammer blows. The wide-striking face and added weight of hand-held sledgehammers are more suitable for securing stakes.

Tape measure

A heavy-duty, inch-wide metal tape measure with a length of 25 feet is just about perfect for most concrete forming jobs. This type of wide metal blade extends at least 7 feet in midair without bending or bowing, an ideal attribute when measuring distances where there is no means to hook the blade's end clip. For extra long walkways measuring more than 25 feet, consider using a 50- or 100-foot tape measure.

Saws

An ordinary handsaw is generally adequate for most forming needs. At the most, you should only have to cut the ends off a few forms to make everything fit properly. If desired, you can even cut the tops off wood stakes to make your screeding job progress a little smoother. Be sure the saw you use is sharp. Dull saw blades make difficult work of cutting straight lines.

Square and string

Unless your concrete design consists completely of curves and rounded features, you'll need a square to make sure corners are set at 90 degrees and forms come away straight from structures such as your house or garage. A small square is also useful as a guide for making straight lines on boards for cutting guides. However, most professionals prefer large squares to accurately gauge corner dimensions and other perpendicular attributes.

Heavy-duty string is a valuable asset in concrete forming, especially while forming long walkways. String is stretched taut between two secured stakes to serve as a perfectly straight guide for installing form lumber. String can also be used to help set up a forming job before forms are actually placed on the ground. By systematically placing stakes and stretching string between them, you can get an excellent idea of how a concrete design will look and function in relation to the existing landscape.

Levels

A string level is an inexpensive and handy tool that can be attached to a section of heavy-duty string or line. (See FIG. 4-1.) Used in conjunction with a tightly stretched string that serves as a form guide, an attached string level shows grade slope. The string can be raised or lowered at either end until it perfectly marks the point where top form edges should be positioned. A taut string signifies a straight line for form installation as well as marking its height position.

Standard carpenter levels are used to set grades for medium length forms measuring up to about 20 feet while a string level is used for walkways in excess of 20 feet. A 2- to 4-foot-long carpenter's level would be better suited for forming patio slabs, driveways, and other small- to medium-sized jobs. Professional concrete form installers rely on 4-foot levels for more accurate readings. This is especially important when forming house slabs, stem walls, and other construction projects. Professionals then use transits for precise measurements and level readings.

When forming small to medium slabs, use, as a minimum, a 2-foot-long level. Although small torpedo levels might be fine for really small work around the house, they are much too short to offer viable readings when working with long form lumber.

4-1 A carpenter's string level helps form installers establish grade. Be sure string is stretched tight to ensure accurate carpenter's string level readings.

Chalk line

A chalk line, also called a snap line, is nothing more than a roll of heavy-duty string secured inside a small metal case filled with regular, powdered chalk. It is used to mark straight lines along house foundations, walls, and other stationery items. A chalk line and powdered chalk are available at lumberyards, hardware stores, and other tool outlets.

With the end of the string secured on a nail or held tight by a helper, you can stretch from one designated marking point to another and then gently lay the string down until it touches the surface to be marked. Then, simply pick up the string, pull it an inch or two away from the surface, and let it snap back. The result will be a perfectly straight chalk line that can be used as a guide for forms.

Other tools

Two miscellaneous items you might also consider purchasing are a heavy-duty carpenter pencil and a nail bag. Heavy-duty carpenter pencils are rugged and hold up well in construction work. The thick, wide pencil lead is strong enough to hold up without breaking while marking lumber, walls, foundations, and other objects during forming. Unless your concrete job is very small and you have no intention of ever doing another outdoor construction project, you might seriously consider purchasing a nail bag set for nails and tools.

FORMING MATERIALS

Basic concrete forming material consists of 2-×-4 form lumber, wood or steel stakes, and double-headed (duplex) nails. This lumber is strong and its size is just right for normal concrete slab depths. Although steel stakes will virtually last forever, their cost is not justifiable for infrequent do-it-yourself concrete finishers. Wood stakes are economical and common at all lumberyards. Be sure to buy several more than you need, however, because many will split or break during forming operations.

Double-headed nails feature a large head about 1/4 inch ahead of the normal nail head at the end of the shaft. These nails are driven into stakes until the first head makes contact with the surface. When the concrete job is complete, the end head sticks out far enough so that it can easily and quickly be grabbed by the claw on a hammer and pulled out.

When selecting form lumber, look carefully at a board's straightness, the most important facet of form lumber. Crooked forms are difficult to straighten, even when pushed and pulled and secured with stakes. And, if your forms are installed crooked, your walkway or slab edge will be crooked, a true sign of an amateur concrete job. Choose each form board separately and look down its length for any signs of bows, twists, or bends. Be very selective and purchase only those that are as perfectly straight as possible.

Lumber is sold in even lengths: 6, 8, 12, 18, and 20-foot boards. Anything longer has to be special ordered and might not be as easy to work with or as straight as desired. If the dimensions of your slab total an odd number of feet, like 9 feet for example, you will have to buy a 10-foot board. You might be able to form it in such a way that the extra foot does not have to be cut off and the full 10-foot board can be used for another project later on.

The longer a form is, the more apt it is to be warped. A 20-foot board is much more likely to have bows in it than a 12-foot piece. Therefore, if your job calls for a 24-foot form board, use two 12-foot boards instead. They can be secured together with a 2-×-4 brace on their backside to extend a full 24 feet. Because they are relatively short, chances are very good that they'll be straight and easy to work with.

Knots

Knots on the sides and edges of form lumber can cause imperfections in concrete finishes (FIG. 4-2). Concrete flows into recessed knot holes and then hardens. Don't purchase boards riddled with knots and deep gouges. If form lumber is perfectly straight with just a few knot holes or gouges on just one side or along only one edge, however, you can use it as long as the bad side or edge is positioned away from the concrete.

Large knots that encompass an entire width section create weak spots in a board. Too much pressure exerted against these wide knots will cause boards to crack and separate. If at all possible, hand select your form lumber and avoid boards with large knots, deep gouges, and other obvious imperfections.

Stakes

Inexpensive wood stakes are sufficient for most jobs and can be found at lumberyards or garden centers. They are sold in bundles of 12- to 24-inches long. It is recommended that short stakes be used in very hard ground, like hardpan, and longer, 24-inch stakes for softer soils. In very hard dirt, you might have a difficult time driving stakes more than a few inches deep. In this case, you can augment form support by pushing dirt up against them. In soft soil, longer stakes will easily be driven deep into

4-2 Knots on form lumber, like this one, are weak spots. Be sure to place a stake next to them for added support. Open knot areas on the 1½-inch side of form lumber must be placed on the ground with only the good 1½-inch side facing up to give concrete edges full form support.

the ground. This is necessary because soft soil will not support stakes that are only partially driven in.

Have plenty of stakes on hand for any concrete forming job. Although your forms will only be holding back about 4 inches of concrete, you would be amazed at the amount of pressure they must hold up against. Once a form breaks and concrete begins to flow out of the break, you will have to stop all pouring operations until repairs are completed. This costs valuable time and quite possibly extra standby fees.

Be prepared to place a stake every 3 feet or so along forms to guarantee that none bow, are pushed out of position, or break. Never place stakes any more than 4 feet apart, regardless of the circumstances. If a wood stake splits on the hammer blow, plan to back it up with another stake or simply place another one alongside it.

In areas where the ground is exceptionally hard or rocky, you might not be able to use wood stakes. Because of the soil and rock conditions, wood stakes might break every time you try to drive them in. In that case, you might have to purchase steel stakes, rent them from a rental yard, or make do with sections of rebar cut to stake size.

Steel concrete stakes can cost from $2.50 to as much as $5 each or more, so in lieu of purchasing expensive steel stakes for just a single concrete job, consider renting as many as you need for as long as you'll need them, perhaps two or three days.

If you cannot locate stakes to rent and new ones are not within your budget, you might be able to adequately support forms by using rebar cut to sections of 1- to 2-foot increments. Although you will not be able to put a nail through rebar stakes, you can nail next to them and bend nails over their shaft for support. Rebar is sold at lumberyards and other construction supply outlets. It is the type of rod you see sticking out of chimneys and block walls as they are being constructed. Rebar is also widely used as an inner grid-work support for concrete retaining walls and other concrete structures.

EXPANSION JOINTS

Expansion joints are complete separations in concrete walkways and other flat work where a type of compressible material fills the gap to absorb the pressure of concrete expansion and contraction. Concrete expands during hot weather and contracts during cold weather. If flat work is confined on all four sides without some type of expansion release, it will crack. Patio slabs and other flat work open on two or three sides are not confined and can generally expand without cracking.

Expansion joints are usually placed about 10 feet apart and are most common in walkways and sidewalks. Although walkways can easily expand sideways, they are definitely confined lengthwise. Expansion joints are also placed at every walkway corner and where walkways butt up against other slabs, such as driveways. If expansion joints are not installed between separate slabs every 10 feet or so along walkways and at every point where walkway corners or bends are featured, cracks will soon appear.

Expansion joint felt

Expansion joint felt is a specifically designed material made of a semirigid felt board treated with a petroleum sealer (FIG. 4-3). Just about every city or town requires felt be used whenever a private slab or walkway butts next to a public sidewalk or other concrete slab. Although the material works excellent for expansion joints, it is rather difficult to work with.

4-3 Expansion joint felt comes in various widths. This one is 3 1/2 inches wide. A bundle of rebar reinforcing rod is resting on the ground under the felt strip.

Felt pieces are placed inside forms before an actual pour, then braced with either a 1×4 or 2×4 to support them perpendicularly. After concrete has been poured around the felt, the braces are pulled out.

Using an edging tool next to felt presents a problem. Normally, edgers have a somewhat sharp leading edge that frequently digs into soft felt material that not only makes a mess of the concrete edge but ruins the crisp felt edge. One way to avoid this problem is to place felt strips a little higher than the top of your concrete. You will have to work around it but edging will be much easier. After the concrete has fully cured, about two weeks, use a sharp razor knife to cut felt down to an appropriate level.

Felt alternatives

If expansion joint felt is not specifically required for your concrete slab or walkway, you might consider using an alternative material that is easier to work with and might be more eye-appealing.

A clear piece of 2-×-4 redwood or pressure-treated wood will serve as an adequate expansion joint. It is easier to install and you will not have to lift it out of the wet concrete as you would a brace for felt, it is easier to edge against, and it might look better once it is painted or stained to match the color of your house. Only redwood or pressure-treated material can be used. Untreated wood will rot in a very short time and be almost impossible to replace.

Another easy-to-install alternative is redwood benderboard. This material is common around flower beds as a border to separate grass from other landscaping materials. It comes in lengths of up to 10 feet, widths of up to 4 inches, and is about ³/₈-inch thick. As opposed to bracing an entire 4-inch-wide piece inside a walkway form, you can simply cut to whatever length is needed and then split benderboard into 2-inch strips. After concrete has been poured, tamped, and floated, wriggle the 2-inch section into place as marked by nails or other obvious designations. You might have to lightly tap it into place. Afterward, a hand float or trowel is used to smooth concrete around the strip, and an edger maneuvered to form a pleasant rounded corner next to it.

This 2-inch strip of benderboard will not completely separate concrete but it will absorb expansion movement to the point where concrete will crack under it but not on top. This option is quick and easy to use on walkways up to 3 feet wide where underlying cracks will not be noticed or cause problems; like when walkway sides are covered by dirt or lawns. Walkways greater than 3 feet in width are too wide for the acceptable use of benderboard strips, as the wood is too flimsy at that length to be able to be inserted in a straight line. Use instead either a section of felt or treated 2-×-4 material.

OTHER FORMING MATERIALS

Double-headed duplex nails are mainstays for concrete form installers. The second head makes pulling these nails from forms a snap. Nails are available in various sizes—6d, 8d, 10d, and sometimes larger. Eight penny, or 8d, is well suited for most common 2-×-4 forming work. They are long enough to go through wood stakes and almost completely through the sides of 2-×-4 forms to offer plenty of strength and stability. 6d nails are a little too small to be used with wood stakes and 2-×-4 lumber, although they might be fine for relatively thinner steel stakes and 2×4s. 10d nails are just too big for most regular forming chores.

For ordinary forming work, nails are first driven through stakes and then into form boards. This is because stakes are placed along the outside face of form lumber. If these stakes were placed and then nailed from inside the forms, you would not be able to remove them once concrete hardened. One exception is when you must nail from the inside using 2-×-4 forms, such as when you are trying to secure a form located on top of a piece of existing concrete.

For example, let's say you want to pour a slab off the back of your house. A sidewalk runs along the side and will be next to the new slab, but you want the new slab to be 4 inches higher than the walkway. How will you place a 2-×-4 form on top of the existing walkway and secure it? Place stakes right next to the walkway—the exact place the side of the new slab will be. Place a 2×4 next to the stakes and on the outside (the stakes will actually be inside the forms). With the form in place, drive a 16d sinker nail through the stakes and through the form and then bend them over on the back side of the form (FIG. 4-4).

The bent nails will hold the form in place. After concrete has cured enough to support final troweling, straighten the nails. Gently pull the form away from the stakes. The stakes will be embedded in concrete on three sides. The only side bare will be the one against the form. Pull the stakes out of the concrete and repair the areas exposed by the pulled stakes.

Drainpipes

Plastic drainpipe is a very good material for directing water runoff. During the forming process, provisions should be made to insert this pipe where needed. Flower beds located between walkways and house foundations could flood during heavy rainstorms. Rigid or flexible plastic drainpipe installed under concrete as an escape for water accumulation would prevent this.

Drainpipe can also be used to redirect rainwater runoff that is blocked from its natural path, such as a new storage shed slab between your house and the fence line. How will water, which normally drains from back to front, get through the slab to the street? If this is not accounted for, water could puddle next to the slab, possibly flooding the storage area. By simply placing a trap next to the slab with a piece of drainpipe attached, water would easily be able to flow away.

Plastic drainpipe is inexpensive, long lasting, and readily available at lumberyards, hardware stores, and garden centers. Be sure it is completely buried beneath concrete so that it does not extend into the slab or walkway. If positioned high into the flat work, a weak spot will develop

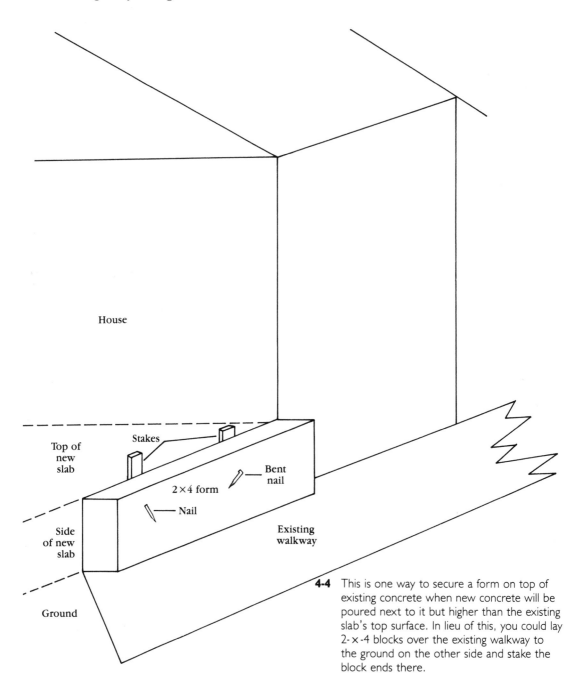

House

Top of
new
slab

Stakes

2×4 form

Bent
nail

Nail

Side
of new
slab

Existing
walkway

Ground

4-4 This is one way to secure a form on top of existing concrete when new concrete will be poured next to it but higher than the existing slab's top surface. In lieu of this, you could lay 2-x-4 blocks over the existing walkway to the ground on the other side and stake the block ends there.

that could lead to cracks because the actual concrete will not be its pre-scribed 3½- to 4-inches deep at that spot. Dig an adequate trench for the pipe. Be sure that trap connections are tight and secure and the outlet opening is clear.

Chapter **5**

Pouring and finishing tools and materials

*T*he equipment needed for pouring and finishing concrete ranges from ordinary sturdy shovels and rakes to specialized items like tamps (jitterbugs) or fresnos (3-foot-wide steel trowel with a handle). Although specific, heavy-duty concrete tools will generally last a lifetime, they need not be purchased. Most rental yards keep plenty in stock and loan them out for very reasonable fees. You can purchase your own set of concrete pouring and finishing tools from a company that specializes in such equipment or at some lumberyards and hardware stores. Prices vary, so shop around (FIG. 5-1).

RAKE AND SHOVEL

Because concrete is such a heavy material, flimsy garden rakes and shovels will not hold up. Plan to use only heavy-duty tools built to withstand construction-type work. You will need a sturdy rake to push and pull concrete toward and away from screed boards during pours. Shovels can be used for this chore, but rakes move more material faster.

Square-point shovels are used for a number of concrete pouring jobs. Although round-point shovels do a good job of scraping concrete out of truck delivery chutes or holding it in place along chutes, they allow too much excess concrete to spill out when trying to place concrete inside narrow forms. Square-point shovels work well to not only place concrete inside narrow forms, but to scoop concrete from the ground and fill in low spots.

WHEELBARROW

Using a wheelbarrow to deliver concrete from a delivery truck to forms is hard work. To help this tough job roll along as thoroughly as possible, use

5-1 Tools needed to finish concrete: (A) round-point shovel, (B) rake, (C) square-point shovel, (D) [optional] power finishing machine, (E) knee boards, (F) fox tail broom, (G) tamp, (H) walking edger, (I) hand edger, (J) seamer, (K) rubber gloves, (L) hand float, (M) pool trowel, (N) scrubber and leaner trowel, (O) finishing trowel, (P) rounded fresno, (Q) bull float, (R) extension poles.

only heavy-duty contractor-type wheelbarrows. Smaller, garden-variety types with solid rubber wheels are not designed for concrete loads. Filled with "mud," they are very difficult to maneuver and tend to spill material while being "wheeled."

Construction wheelbarrows are reinforced with metal uprights, have thick, long wood handles for strength and maneuverability, and are designed to move about without spilling loads. Construction wheelbarrows are also built with a braced front nose piece that functions as a stop against the ground when they are tipped upright to dump loads over forms and into slab areas (FIG. 5-2). They are readily available at most rental yards. New ones, purchased from a lumberyard or concrete supply outlet can cost up to $100.

5-2 A heavy-duty contractor's wheelbarrow with an air-filled tire is the best for concrete work. Long handles and a sturdy front nose piece help "wheelers" maneuver and dump concrete loads confidently.

If you have never wheeled concrete before, you must start out with your wheelbarrow only filled part way; no more than half its capacity, even with construction-type wheelbarrows. Although it may not seem like much material, you will be surprised at how heavy the load is and how much physical effort is needed to push the wheelbarrow, turn it around corners, and then tip it up to empty. As you progress through your job, expect to gradually have the concrete truck driver fill your wheelbarrow more and more as you get accustomed to the work. Never fill it to the brim; it is just too heavy to manage.

SCREED BOARDS

Screed boards are straight, 2-×-4s used to level fresh concrete just after it has been placed inside forms. The bottom edge of a screed board is usually laid on top of forms and then pulled across them to sweep back piles of concrete sticking above forms. They are also used to find low spots where more concrete is needed to fill up a form. In essence, a screed board's bottom side determines a slab's top surface level.

Concrete finishers operate screed boards in a sawing-like motion across the tops of forms. This action works well to help aggregate (rock) settle into the surface mixture and move material back while flattening surfaces to a level consistency.

Screeding maneuvers are hard work, perhaps the physically most demanding part of any concrete job. It is also one of the most integral parts of the entire concrete finishing process. Poorly screeded slabs and walkways frequently have uneven surfaces riddled with low spots that hold puddles of water and look terrible.

By all means, take whatever time necessary to ensure perfect screeding during all parts of your pour. If necessary, stop everything else and make the concrete truck sit idle until your screed workers catch up with the concrete that has already been dumped. Should you decide to ignore the importance of screeding and try to catch up later, you'll find that an enormous amount of work will be needed to move excess concrete from high spots and then over to the low ones. While you are busy doing that, concrete will be setting up and you will lag further and further behind on other work that needs to be done.

Most professional concrete finishers realize the hard work needed to accomplish first-rate screed jobs. They also feel that once screeding has been completed, the majority of work is over; at least the hard work that demands most of their physical exertion.

TAMP

Tamps are metal tools commonly referred to as *jitterbugs*. They consist of a metal screen mesh that measures 3-feet-long on the bottom and about 8 inches in width. Attached to each end of the bottom screen are metal arms that reach up to a connecting bar featuring two handles. This tool is used to tamp (push down) rock and aggregate on concrete surfaces in order to bring up creamy textures suitable for finishing. Slabs that are not tamped will have extra rough and rocky surfaces and lack the amount of necessary cream for smooth finishing endeavors.

Most concrete jobs only require tamps be raised a few inches off the surface and then allowed to fall down flat onto the mud. A lot of forceful downward pressure is not necessary. The screen size is just right to force rock down and allow creamy mud to filter through. Results should be relatively smooth surfaces with rocks pushed down at least $1/2$ to 1 inch below a layer of cream. Tamps leave their own pattern over an entire slab surface but they can be quickly and easily wiped out with a bull float.

Tamping can be a very messy chore. Each time the tool comes in contact with fresh concrete, cream will splatter. Expect your boots and pants to get covered. When pouring next to, or near, your house, garage, fence, or other structure, be sure to erect some type of a shield against concrete splatter. Before pouring begins, use strong duct tape or heavy-duty masking tape to secure plastic or newspaper against walls and other things you want protected. Generally, you only need to place protection up about 3 feet from the slab's surface. Should tamping splatter occur higher than that, you are tamping much too hard.

Be sure that newspaper or plastic is not positioned so low as to interfere with pouring or screeding operations or that it actually comes in contact with a slab's surface. The protective material you put up can be removed while you are out on the slab doing the finish trowel work. At that time, should any concrete residue fall off the newspaper or plastic, it can be easily trowelled into the surface.

BULL FLOAT

Screen mesh patterns left behind by tamping are smoothed with a bull float. Although metal bull floats are available, many professional concrete finishers use bull floats made of wood. Essentially, bull floats are 3-foot-long boards, about 8 inches wide and 1 inch thick, equipped with a metal support and a swivel connector to which extension poles are attached because you cannot walk on slabs at this point (FIG. 5-3).

Extensions (poles) are attached to bull floats to reach hard-to-reach places without having to step in concrete. Extensions are usually 6-foot-long poles with locking mechanisms which allow them to connect to the

5-3 Bull float with adapter and extension poles.

bull float adapter and also lock into each other. For a slab measuring 18 feet, you would connect one 6 foot pole to the bull float and then two more in succession to the first pole for an overall reaching span of 18 feet.

A bull float is pushed across and then pulled back over wet concrete slabs to remove tamping marks and generally smooth surfaces to an even and glassy texture. This tool is also used from various angles to fill in cat's eyes—small and shallow low spots easily noticed on slabs that the bull float never touched. Maneuvered from different angles, bull floats can pick up cream from surrounding areas to help fill in shallow cat's eyes.

The swivel feature on a bull float's extension pole is used to position extensions in such a way that the float can be pushed out onto a slab surface with its front edge lifted slightly. This prevents the tool from digging into the concrete.

FRESNO

The fresno is a finishing tool that looks and operates much like a bull float. The only difference is that instead of an operating surface made of wood, a fresno features a metal blade (FIG. 5-4). Its purpose is to continue smoothing concrete as it stiffens and becomes harder. The extension poles used for a bull float are the same used for a fresno.

5-4 The rectangular fresno on the top with square corners often leaves lines on fresh concrete. Fresnos with round corners are more line-free. Note the adapter on the round fresno, which can be adjusted a number of ways.

Fresno blades are 3 feet long and about 6 to 8 inches wide. They come in two styles; one with a rectangular shape and square corners and another with rounded edges and no square corners. As with a rectangular bull float, square-edged fresno blades tend to leave lines behind after being operated on fresh concrete. The rounded model, on the other hand, tends to be line-free and is a preferred option.

Fresno use is determined by concrete surface stiffness. Used on concrete that is still very wet, a fresno will not accomplish much more than turning a top layer of cream into a soupy mix. Novice concrete finishers tend to work concrete too much. By doing this, they continually agitate the top layer of cream until it becomes saturated with sand and grit. As a general rule of thumb, a concrete slab that has been poured relatively wet should only be bull floated a maximum of two times and smoothed with a fresno about three times. The use of these tools are covered in more detail in chapters 10 and 11.

HAND FLOATS AND TROWELS

A hand float is basically a piece of flat, smooth wood with an attached wooden handle measuring about 16 inches long and 3 inches wide; larger and smaller sizes are available. Hand floats are used to help place, move around, and smooth wet concrete as it is being poured into tight spots, against walls, and along edges. Like a bull float, they are used to flatten concrete while it is still wet and easily workable. After concrete starts to stiffen, wood floats are cleaned and put away, making way for metal fresno tools and metal hand trowels.

Hand trowels are metal blades equipped with handles. They range in size from about 12 inches long and 3 inches wide to 18 inches long and 4 inches wide. Most professionals prefer to use a small "scrubber" hand trowel (14 inches long and 3½ inches wide) to work up a layer of cream on a rather rough surface and then finish it off smooth with a "cheater" trowel that might measure 16 inches long and 4 inches wide. They prefer this method because it is easier to quickly and vigorously operate a smaller trowel for scrubbing and then simply wipe it smooth with a swipe or two from a larger trowel.

The blades on metal hand trowels must be in excellent condition to smooth concrete without leaving behind lines and other imperfections. Nicks and dents along a trowel's edge will scar concrete surfaces and make the finishing effort a difficult one. When renting hand trowels, be sure to inspect the bottom side and all four edges. Should you detect imperfections, ask to rent a different tool; or have the rental yard service department file the edges clean and even.

As a knife blade becomes sharper as it is slid over a honing stone over and over again, continuously operating a metal hand trowel over concrete causes its edges to sharpen too. Never run your hand or finger across the edge of any hand trowel. Like a razor, these edges can quickly slice through skin.

You'll need at least two metal hand trowels, one smaller than the other, to finish almost any concrete slab. Along with working up cream with the small one and actually finishing with the larger (scrubbing and cheating), you'll normally have to lean on one trowel in order to reach spots in the middle of a slab with the other. Unless your concrete job is only a narrow walkway, you'll find yourself leaning on one trowel to reach even further to finish areas away from the edges.

Pool trowels function exactly the same as metal hand trowels except that the front and back ends are rounded. As with a round fresno, pool trowels work much better to finish concrete because they will not leave lines behind. If possible, rent at least one rounded pool trowel to work as a cheater. Should you want to purchase your own, be sure to use the term *pool trowel* because these tools were originally designed to finish the mortar and plaster used to coat in-ground swimming pool sides, which are seldom perfectly flat all around.

Professional concrete finishers usually carry their hand tools in a large, five-gallon plastic bucket. Besides the convenient size for carrying tools to and from work, these buckets are partially filled with water at job sites and used to clean trowels after each use. After rinsing off all concrete, trowels are hung over the edge of the bucket to dry. **CAUTION**: be extremely careful washing metal hand trowels with your hands. Their edges can quickly cut you. Consider using a small brush with rather soft bristles.

KNEE BOARDS

Because the final finish on concrete slabs is achieved by use of metal hand trowels, you are probably wondering how the center part of a slab measuring 14 feet or more is reached. Knee boards are the answer. Knee boards, which generally measure about 1¹/2 to 2-feet square, spread finishers weight over a wide area (FIG. 5-5). Because concrete cannot be hand finished until it has reached a fairly hard texture, concrete finishers rely on these small pieces of smooth plywood to kneel and walk on while maneuvering over large slabs. By the time concrete sets up enough to support finishing, it is also hard enough to support the weight of finishers.

Finishers always use two knee boards. This way, they can leapfrog around slabs by placing one down and standing on it while positioning the other, and so on. Finishers start in a corner where the first mud was poured and then move in a parallel direction to how the slab was

5-5 Knee boards are required for all slabs that are too big to reach from outside forms. This one has a small piece of wood nailed to one side to use as a handle.

screeded. Essentially starting on the concrete that was poured first and ending up in the area where the last concrete was poured.

Knee boards *do* leave impressions on concrete, however. But as long as the wood on their bottom surface is smooth and not pockmarked with knots and gouges, imperfections can be quickly and easily wiped out. A scrubber is used to roughen the area and bring up cream, which is then wiped out with a cheater to a smooth and line-free finish.

If you were to simply walk and kneel on a slab to finish it, your feet would leave deep impressions, knees would dig in, and toes would severely scratch and mar the surface. On that note, be sure your feet rest on the knee board positioned behind you as you kneel on the other board during finishing. Also, because slightly wet concrete can cause knee boards to stick to slab surfaces quite strongly, many finishers nail a small strip of wood to one edge for use as a handle.

Because knee boards leave impressions on concrete surfaces, always try to finish as much concrete as possible from each knee board site. Reach out as far as possible while leaning on one trowel. Finish areas in front of and on both sides of knee boards before moving on. In addition, be certain that knee boards are not spread out too far from each other. Ideally, they should be set in such a way that your knees easily fit on one while your toes can rest on the other. Work your way from front to back or side to side around the slab. Once you reach a form, step off to complete that section; then start a new path. Knock any mud or debris off your feet before stepping on knee boards to prevent that debris from accidentally falling onto the slab surface.

EDGER

Concrete "edges" refer to the very top corner of concrete slab sides; the 90 degree corner produced at those points where a slab surface stops and the actual 4 inch side of the slab begins. Essentially, edges are those bordering points where the top portion of concrete slabs touch forms. By looking at almost any existing concrete sidewalk or patio slab, you should notice that these edges are slightly rounded off, instead of being perfectly formed at a sharp, 90-degree angle.

Rounded concrete edges are purposely made with a hand or walking edger (FIG. 5-6). Creating rounded edges is this tool's sole function. Similar to a metal hand trowel, edgers feature a distinctive, rounded-off side. It is this side that is introduced between tops of forms and concrete. As it is smoothly maneuvered along outer slab perimeters, concrete cream is shaped to conform with the curved arc of the edger tool. Although 3/8-inch arcs seem to be the most popular, these arcs range in size from a tight 1/4 inch to a full 1/2 inch.

Walking edgers are almost identical to hand-held models except that they have long handles. Instead of having to squat or kneel down to use a hand-held edger, walking edgers allow finishers to walk along slab perimeters to effect rounded edges. Walking edgers are also almost always larger than hand-held tools. Although the larger size creates identical arcs, it is

5-6 The walking edger on the left and the hand edger on the right have identical rounded arc sides.

easier to produce crisp edges while finishing a wider swath along slab borders. When either purchasing or renting a combined set of edgers, be certain the arc for the hand-held tool matches the arc for the walking one.

To make the job of concrete finishing a little easier, many professionals like to lightly edge concrete soon after applying a bull float. Although concrete is still much too wet to support a crisp edge, all of the aggregate is pushed down far enough so that future edging maneuvers will not kick up aggregate obstacles. As fresno operations continue, more and more cream is deposited along edges to fill in low spots that might have been produced by early edging. This is good because it provides an excellent surface for edgers to work on later while affecting final edging maneuvers.

SEAMER

Along with expansion joints, concrete finishers install control joints (seams) about every 4 feet along walkways, driveways, and at other crack-hazard points. For the most part, control joints are deep grooves put into concrete finishes in anticipation that if cracks do occur, they will follow seams and not randomly appear anywhere else. Along with expansion joints, seams help to control concrete cracking as much as possible.

Seamer tools are made of metal and feature a raised ridge down the middle of their face (FIG. 5-7). To insert seams, finishers lay a straight 2-×-4 board across a freshly bull-floated slab to rest on forms at both sides of the concrete. The correctly positioned 2×4 is then used as a guide to run a seamer against while forcing it into the concrete. The result should be a definitive line across concrete where rocks and aggregate are pushed down and cream is smoothed.

As with initial early edging, don't expect that first seam to stay perfectly clean and crisp after maneuvering a fresno over it. Cream will surely fill the groove to some degree. Because aggregate has been forced down and replaced with just cream, however, you should have no problem

5-7 This hand-held seamer features a raised ridge in the middle of its bottom side that actually makes grooves in wet concrete.

refinishing seams at a later stage. In fact, with a steady hand or use of a walking seamer, you might not even have to use a 2-×-4 guide again.

It is common to place seams every 3 to 4 feet along walkways with an expansion joint every 9 to 12 feet, whichever works best proportionately. Try to maintain evenly spaced seams and expansion joints, not only to prevent uncontrolled cracks, but to look visually attractive. For example, if your walkway will be 24 feet long, plan on installing an expansion joint in the center (at the 12-foot mark) and then put in seams at the 4, 8, 16, and 20-foot marks.

Also, plan on installing expansion joints or a combination expansion joint and seam where walkways or other slabs form a corner so the corner is outlined in a square pattern. An example would be a walkway that comes off a patio slab located at the back corner of your house and then goes down along the side toward your front yard. The point where the walkway meets the slab should have an expansion joint and the point at which a line would designate a 90-degree angle with the expansion joint should have a seam. In essence, the seam would be straight in line with the back of your house to make the first part of the walkway squared off—expansion joint on one side, seam on the other, and lawn on the two open sides.

BROOM

The most common finish for exterior concrete flat work is called a broom finish. It is very easy to put on and hides a multitude of minor hand trowel flaws. It also provides concrete with much-needed traction qualities when wet. Smooth-finished concrete is as slippery as ice when wet. Have you ever noticed how slippery your garage floor is after parking a rain-soaked car on it? Outdoor concrete must be broomed, or finished, in ways other than simple trowel finishing if you expect to be able to walk on it safely during wet weather or any other time the surface is wet—like when hosing it off or while watering the lawn.

Finishers use a soft-bristled push broom or one that is specifically designed for concrete work. Stout-bristled brooms frequently used to sweep asphalt and other coarse surfaces are much too strong and will leave very deep and uneven lines in concrete. Soft push brooms work well to effect attractive lines that not only look good and cover small imperfections, but also cover wide swaths with each pass.

In most cases, the tiny lines left behind as a result of pushing or pulling a soft-bristled push broom across a finished slab scratch the surface just enough to allow adequate traction. If you are pouring a steep driveway and anticipate slippery winter weather conditions, consider using a strong-bristled broom to effect heavy, deep lines to really aid traction.

To reach parts of a slab located further away than your push broom handle will allow, place an extension pole over the broom handle with the featured hole of the extension pole positioned closest to the broom bristles. Through that hole, insert a small wood screw into the broom's wooden handle. This should secure the extension to the broom so that it cannot be pulled off (FIG. 5-8). To that extension, you can attach additional extensions just like you did for the bull float and fresno. If the slab is still too long, or obstacles like fences are in your way, you can operate the broom from opposite sides of the slab to create the broom-finish effect by brooming half the slab from one side and the second half from the other side.

Broom lines do not have to be necessarily straight. Wriggling the broom as you pull it toward you creates a wavy broom design. You could also use a small foxtail broom to create a pattern of small semi-circles. Get back on the slab with your knee boards and simply operate the foxtail

5-8 A soft-bristled push broom has been reinforced with brackets on each side, and a screw in the center of the handle close to the broom head is used to secure extension poles.

broom in "half moon" arcs until you achieve the finish desired. The final design will look something like a windshield wiper pattern with lots of small lines following a series of half circle patterns.

Concrete must be set up and trowel finished before a broom is maneuvered over it. If your concrete has set up so much that a soft-bristled broom will not make any lines, try putting pressure on the bristles by pushing down on the broom handle while pushing and pulling it across the slab. If that doesn't work, dip the bristles in a bucket of water so they are soaking wet and then apply them to the surface. Place it as far away as you can and then just pull it toward you. In extreme cases, you will probably have to both push and pull it across the slab.

PROTECTING SURFACES

Concrete work can be very messy. No matter how careful you are, concrete usually gets splattered on something—usually the walls along your house. To prevent unnecessary concrete splatter on the house or other structures, use heavy-duty duct tape to secure newspaper or plastic on walls before pours. Plan to cover an area that reaches from the concrete's top surface to at least 3 feet above. This should protect against splattering from wheelbarrow dumping and tamping maneuvers. Newspapers or plastic are removed while out on slabs with knee boards hand finishing. Helpers may be needed to assist in getting splatter protection away from slab areas. Any concrete residue that falls off onto slabs is finished into surfaces with a hand trowel.

Do not place newspaper or plastic so low that its bottom edge is allowed to touch a concrete surface. Should it become buried in concrete, a slab's finish will be marred when you attempt to pull it out. In addition, should a lower section break loose from a wind gust, its flapping action will make all sorts of marks on any slab surface it touches. If need be, chalk a line an inch or so above proposed slab surfaces to designate the lowest point splatter protection material can be placed.

Besides house walls, there may be other items you need to protect against concrete splatter. Railings, metal patio cover posts, sliding glass door sills and pool pumps are just a few. Heavy-duty duct tape may be wrapped around items or used to secure strips of plastic in place. If concrete does find its way to an unprotected surface, clean it off as soon as possible with a wet rag or sponge. Once it hardens, concrete is difficult to remove and may require use of a wire brush.

RUBBER BOOTS AND GLOVES

The cement powder used to mix concrete is a powerful agent. Not only will it tend to dry out and otherwise ruin normal shoes, it can severely dry out the skin on your hands, resulting in cracks, flaking skin, and other problems. Professional concrete finishers always have a pair of heavy-duty rubber boots available for pouring concrete. Worn during pouring and tamping exercises, they help to keep their feet dry and hold up against the potent effects of cement in concrete. When the work that calls for them to walk around in concrete has been concluded, finishers will usually change from rubber boots into more comfortable leather work boots or other suitable shoes. When the slab has been completely finished and forms pulled, they might put on the boots again while using plenty of water for tool, form, and equipment cleaning.

Plan to wear a pair of heavy-duty rubber gloves while pouring concrete. You will often find yourself reaching into the mud to pull out handfuls of concrete in order to fill tight spots or move it away from the screed board. The lye in concrete dries out the skin on your hands quickly unless they are protected with rubber gloves. Many professionals are forced to wear gloves not only to keep their hands in decent shape, but also to prevent the onset of concrete poisoning. Concrete poisoning is a result of continual exposure to the powerful substances in concrete and causes an infection along hands and arms.

Personal safety during any do-it-yourself project is a major concern. You have chosen to pour your own concrete job to save money and to enjoy the satisfaction of a personal achievement. Do not take any chances on ruining that experience by avoiding the use of such inexpensive safety equipment as rubber boots and gloves. For that matter, plan on wearing eye protection, too. Even plastic sunglasses offer protection against accidental concrete splatter in your eyes. Should this happen, however, thoroughly flush your eyes with plenty of clean water.

Chapter **6**

Site preparation

*P*reparing an area for concrete work involves a number of chores. First, remove obstacles from the immediate job site that might hinder overall maneuverability. Patio furniture, swing sets, planter boxes, and the like should all be relocated to a remote part of your yard so that they are totally out of the way. You will need plenty of room to operate long bull float and fresno extensions. Site preparations might also include removing old concrete, relocating sodded lawn sections, transplanting flowers and shrubs, pulling old tree stumps, or digging a desired grade. Systematically plan your site preparation job so that each chore is completed before starting a new one. In other words, don't start haphazardly digging in the middle of a site until you have established the required grade by way of forms, string, and level.

REMOVING OLD CONCRETE

Front doors, sliding glass doors, and other residential door entries often have small concrete steps or stoops at their outdoor entrances. They are usually only big enough to accommodate a small welcome mat and a place where you can remove muddy shoes before entering the home. When planning concrete slabs or walkways in areas surrounding them, plan for their complete removal as opposed to just pouring new concrete around them (FIG. 6-1). Unless the actual size of the steps or stoop fits within planned expansion joint or seam patterns, they will appear awkward and out of place.

Old concrete can be taken out with a jackhammer or busted up with a sledgehammer. For small jobs like stoops, most professionals prefer the latter. With a pick and shovel, remove as much dirt as possible from beneath the outer stoop edges to create an open space under the concrete. Then, force one end of your pick under the stoop to raise its front side off the ground and loosen its grip on the foundation it is positioned

6-1   Rather than pour new slabs around small, existing stoops or landings, remove old concrete so fresh pours look uniform.

6-2   Use a heavy-duty pick or large pry bar to raise the front of old stoops and loosen them from the base. This worker has his foot on a pick's wide blade and is pushing down on it for added leverage.

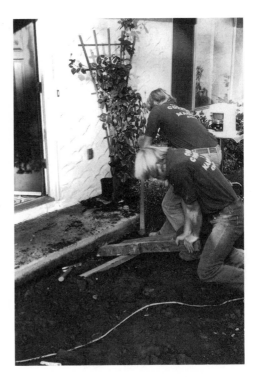

6-3 Some concrete landings are difficult to break loose. Along with being heavy, they sometimes have an adhesive-type bond to walls. Here, workers use a 2-×-4 block and 4-×-4 lever along with a pick to loosen a stubborn front stoop.

against (FIG. 6-2). Use the pick to take advantage of any available leverage. Concrete is very heavy and you might be able to only loosen a little at a time. If need be, establish a better leverage system by using a long, 4-×-4 post as a lever blocked with short 2×4s (FIG. 6-3).

Once the stoop has been broken loose, use the pick to pry it away from the house. Try to slide it straight out horizontally, as upward movement could chip or crack existing plaster or wood siding immediately above it. Moving it out should eliminate any chance of causing damage to the structure while pounding on it with a sledgehammer. Once the old concrete is broken up and removed for disposal (FIG. 6-4), you can begin forming.

FORMING

The quickest and simplest way to start almost any forming job is to lay forms out on the ground in their planned locations (FIG. 6-5). This allows you to better perceive the dimensions of your planned slab or walkway. It also allows you to strategically place each form so that few, if any, have to be cut to fit. In other words, don't waste good lumber. Measure forms and place them in spots where full advantage can be taken of their length instead of cutting good long forms to fit where shorter ones would work fine.

Use forms to their greatest potential. If one side of a slab measures 13 feet, use a 14-foot board and allow it to extend past the corner by a foot

6-4 A tarp protects the front door as concrete is repeatedly smacked with a sledgehammer until a pick is able to pry out concrete chunks in small, manageable pieces.

6-5 Outline proposed jobs with form lumber before pounding stakes. This gives you the opportunity to properly place forms and take full advantage of form sizes without having to unnecessarily cut long forms to fit.

(FIG. 6-6). The extra extension will not cause any problems and you will prevent wasting an otherwise functional 14-foot, 2-×-4 form.

Plan on having scrap lumber on hand that can be cut to size for jobs requiring 1- or 2-foot form boards. If scraps are not available, purchase short 2×4s that you can use to cut out a sufficient number of small forms. Along with preventing waste, preserving long forms means they will be on hand for future concrete flat work projects or to use in another building project.

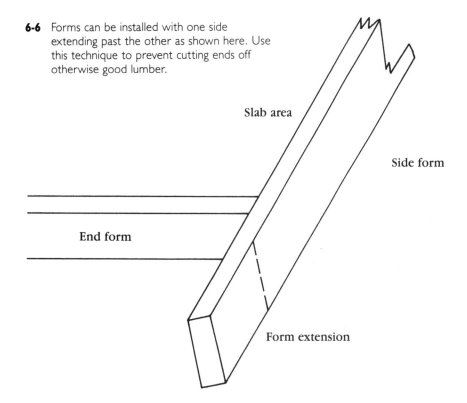

6-6 Forms can be installed with one side extending past the other as shown here. Use this technique to prevent cutting ends off otherwise good lumber.

Slab area

Side form

End form

Form extension

Working with the landscape

In many cases, areas planned for concrete are already landscaped with grass, flower beds, or mulch. Chances are, those areas must be excavated to make room for forms to fit in such a way that their top edges are positioned according to the grade level planned for the concrete. Simply resting on top of a lawn, for example, forms might sit too high in relation to where you want the top surface of concrete to be located.

In this case, lay forms in their relative positions and measure how high they sit with relation to where you want them to. Extending perpendicular from the house, you can measure the distance from the top edge of the form butting against it to where you actually want the top of the form to be. This measurement tells you how much dirt has to be removed in order for the form to fit according to grade. Knowing this, you can accurately judge how much dirt has to be dug out rather than guessing and digging more than necessary.

Further, by having forms laying close to their prescribed positions, you can use them as guides to establish actual digging perimeters (FIG. 6-7), especially when digging out lawns. Using a straight guide helps you to confine excavation efforts to the slab area with minimal disruption to the rest of the landscape. It saves unnecessary digging work, too. Square point shovels are great for this work, as their flat blades cut sod evenly and can scrape out dirt smoothly.

6-7 Once a job is outlined, you'll have a good idea where forms are to be placed. Use a square shovel to cleanly dig out sod or other ground material. A neat job will save extra landscaping renovations later.

Blocking forms together

Slabs and walkways are easiest to form using lumber that has identical dimensions. A 10-×-12-foot slab poured against the back of your house, for example, is most easily formed using one 12-foot and two 10-foot 2×4s. This is great for jobs featuring such measurements, but what about long walkways that are more than 30 feet?

Because locating straight, 2-×-4 form boards in excess of 20 feet is difficult and expensive, you will have much better luck blocking two shorter 2×4s together to make one long straight form. For a 30-foot run, a blocked form consisting of 16- and 14-foot boards is perfect.

Lay the boards down end-to-end on a flat surface. Determine which side of each board you want facing the concrete (their best sides) and which edges you want facing up (their best edge). Put the good side face down and note which edge is face up by marking it with a pencil. This will guarantee that after the boards are blocked together, their best faces will be located against the concrete and their best edges will be on top to make concrete edging easiest.

Drive a nail part way into the far end of each board. Attach a string to one nail and then stretch it tight to the other nail. Secure it with a good knot. This string will serve as a guide to show that both 2-×-4 boards are aligned with each other. In lieu of a string, you could butt another long 2×4 against the two on the ground to ensure they are lined up correctly.

Use a 2 to 4 foot, 2×4 as a block. Center it over the spot where the two 2-×-4 forms butt together. Securely nail it to both forms using at least six to eight nails on each form (FIG. 6-8). While nailing, constantly check your string or 2-×-4 guide to ensure the forms remain in a straight line. This is easiest by completely nailing the block to one form first and then the other. Be sure the block's top edge matches the form lumber's top edge. If it sticks up higher, it could throw off screeding levels. In addition, have someone help you carry blocked forms to their working location to prevent bowing or weakening the block.

6-8 Two long 2-×-4 boards butt next to each other to form part of a driveway pour. A short 2-×-4 block has been nailed to both forms to strengthen the joint and keep it straight.

ESTABLISHING THE HIGH POINT

The high point of any concrete job determines the direction that water will drain. Typically, high points are almost always located next to house foundations or other walls so water will run away from them and toward a yard, drain, or street. Deciding on the correct high point of any slab or walkway is critical. Misplacing high points can cause water runoff to flow toward a house or structure and puddle, eventually causing dry rot, mildew, and other water-related problems.

When pouring concrete next to front or rear doors, use an existing doorsill as a guide (FIG. 6-9). (Always attempt to remain at least 2 inches below floor levels to prevent water, dust, and debris from easily entering through door spaces.) Make pencil marks on the wall at each end of the sill, level with its lowest edge. Measure down from those marks to where you want the top of the concrete to be (FIG. 6-10). Join these two marks

6-9 A carpenter's level is used to designate where a pencil mark should be made to the side of a doorsill. This mark will be used as a high-point guide.

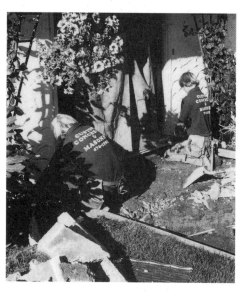

6-10 Pencil marks located an equal distance from each side of the door show where forms should be placed so that the door is centered in the middle of a new concrete slab. Pencil marks are also used to determine form height against the house.

together with a chalk line. Chalk lines are also used to denote grade levels along the sides of structures, such as down the side of a house to show where a walkway surface will be or along the back wall to designate the top of a patio slab (FIG. 6-11).

Before chalking any line, though, grade levels have to be determined at each end and then marked with a pencil. These marks are measured down from a wall's plaster lip, the bottom of a horizontal piece of siding, or any other predominant feature that is installed level.

For a walkway down the side of a house, for example, the mark at the back corner of the house could be located 2 inches down from the wall siding's bottom edge, while the mark at the front corner of the house will

6-11 This chalk line denotes where a slab's top surface should be. This helps while grading and during concrete placement.

be located 4 or more inches down from that siding's edge. A chalk line snapped between these two marks will show that the walkway will run downhill from the back of the house toward the front. Note: snapping long chalk lines might require three people; one at each end to hold the line on their marks and one person in the middle to actually pull the line away from the wall and snap it (FIG. 6-12).

6-12 Snapping long chalk lines against a wall or foundation might require three people. Two on the ends to hold string tight, and one in the middle to snap the line.

SQUARING FORMS

The corners of rectangular slabs and walkways should be formed at right, 90-degree angles. To ensure that forms start out square when pouring patio slabs and other flat work against a house or other structure wall, use a carpenter's square. Position one side of the tool along the affected wall of the structure and adjust the form until it sits perfectly square with the other arm of the tool. Secure that form with stakes and set the next form square with it. Continue around your slab making sure all corners are square before actually securing them to stakes.

Squaring forms for concrete jobs where one side abuts an existing structure, like patios and walkways next to houses, is easy because you have something straight to work off of. Setting up a square slab evenly distanced away from structures entails a little more work and ingenuity. For this type of form squaring job, you'll need a roll of sturdy string line and at least six stakes, along with a carpenter's square. This job is easiest when two people work on it together.

If one side of your new slab is designed to be in line with an exterior wall of your house, place a stake right next to that wall so string can be stretched from it along the wall for about 10 to 12 feet. Now, grab the roll of string and stretch a line to about 2 feet past the farthest side of the proposed slab's edge. Move the string back and forth until it lines up perfectly with the house wall. Place a stake in the ground at that point and secure the string to it. This string should be perfectly square with the corner of your house. Use a carpenter's square to be sure that it is, however, and adjust as needed according to the square tool. Place your first form in line with this string and square up the remaining forms to it.

For slabs positioned more or less in the center of a house wall, where side walls cannot be used as just described, you will have to be a bit more inventive. Initially, place a stake along the wall of the house that parallels the proposed slab and denotes the point at which one of the perpendicular sides of the new slab will be located. Attach a string to it and stretch it out to a point about 2 feet past the proposed slab's farthest side.

Have your helper hold one arm of a carpenter's square against the house wall with the other arm next to the string. He can then signal you when the string is square. Place a stake at the square point and secure the string tightly to it. Check the square again to be sure the string is placed correctly. This string will give you a straight line to use as a guide for setting the first form. Use a carpenter's square to correctly position forms that butt against this first form and then again while setting the form parallel to it (FIG. 6-13).

OTHER TASKS

Sometime before actually starting the forming process, you should prepare and plan for other tasks that will require your attention. Although leveling and securing forms is pretty straightforward, setting up a system for screeding might rely on which direction concrete will be delivered from. Screed work always starts at the point where concrete is first

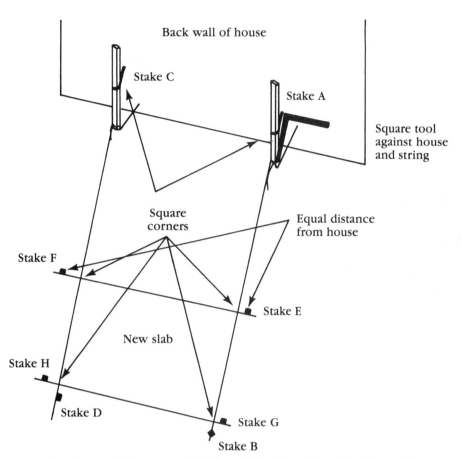

Back wall of house

Stake C

Stake A

Square tool
against house
and string

Square
corners

Equal distance
from house

Stake F

Stake E

New slab

Stake H

Stake D

Stake G

Stake B

6-13 Forming a perfectly square slab in the middle of a yard parallel with an existing
structure without using special equipment can be done with stakes, string, and a
large carpenter's square. The distance between stakes A and C denote slab length.
Stakes E to G and F to H show slab width. This job starts with stake A and runs
alphabetically through H.

placed. Can a concrete truck fit into a space on one side of the slab or will
you have to wheelbarrow concrete from the other? Will you need to build
special ramps up steps or over other obstacles for wheelbarrows en route
from the concrete truck to the work site?

   Is there a plumbing cleanout located inside your proposed slab area?
If so, you will need to keep it free of concrete so that it can be accessed.
Some concrete finishers cover cleanouts with large coffee cans placed
upside down over them with their bottom sides evenly lined up with the
slab surface. When access to the cleanout is needed, the bottom of the
coffee can is simply cut out. These items can also be protected with ordinary lawn sprinkler control boxes, which are quite similar to covers used
to protect residential water meters.

   Horizontal vents located along house foundations must be left clear.

To hold concrete back from them, use 1-×-4 form boards. Are basement window levels located below your proposed slab surface? You will not want to pour concrete against them, so plan form work accordingly.

Will you eventually want to extend a water or electrical line from your house to an area beyond the new slab or walkway? Provide for that possibility by inserting conduit or plastic pipe under formed areas so that the option is always open.

Discarding old concrete and landscape debris might be a problem, especially if you do not own a pickup truck. You will have to arrange a means of disposing of unwanted debris.

Finally, as you prepare to actually start forming, lay out your job in such a way that you will not have to continuously get up and down to retrieve stakes, nails, tools, and the like. With forms placed as designed, lay down a stake every 3 feet or so around the entire perimeter. Use a nail bag filled with plenty of duplex nails and the tools you will be using. This way, once you get a form correctly positioned, you can grab a nearby stake, pound it in, and nail it with a minimum of effort.

Chapter 7

Basic forming techniques

*C*oncrete forms are strategically placed and secured to accomplish one goal—to keep wet concrete confined to a certain space in a predetermined shape until it hardens and can remain stable on its own. This is a simple concept. However, should these forms be placed in an illogical manner without regard to surface height, straightness, or grade, the end result will be a hardened slab of concrete that looks awkward and unprofessional and might very well prove to be functionally inadequate.

Forms must be strongly braced with plenty of stakes to ensure they remain in position while being subjected to the weight and pressure of pouring concrete. In addition, every form installation job must account for water runoff, whether it is to assist in yard drainage or to prevent puddles next to a house or other wall structures.

From a visual perspective, forms must be straight in order to flow with the basic symmetry of any landscape. If forms are not installed straight, there is no way concrete will harden straight. Crooked concrete edges wobbling along the sides of slabs and walkways are signs of unprofessional workmanship and cannot add distinction to any landscape design.

FIRST FORM

Most residential concrete patio slabs and similar jobs are poured next to houses or garages, therefore, the following examples depict common designs in which one wall of a house or other structure serves as one side of the slab's forming needs. In other words, concrete will be poured against the bottom of the wall to make it, in essence, one of four required forms. Therefore, the first form you install will actually be a side form.

First, mark the structure wall to designate slab height, as described earlier—at least 2 inches below the bottom of a door sill, house siding board, wall plaster lip, etc. Be certain it is marked at the spot you intend to place your first side form.

Butt the end of your form to the wall so that its top edge is located on the mark. The form should stick out perpendicularly from the wall at a right angle. Check it with a carpenter's square. If the grade is too deep for the form's top edge to meet that mark, push a little dirt under the form to raise its edge. If the grade is too high to allow the form's top edge to go low enough to meet the mark, use a square point shovel to scrape away dirt. Adjust the grade along this form's length until it rests in a position where its top edge is in line with the mark.

Once this form has been closely adjusted to grade and angle, position yourself inside the forming area and pound a stake into the ground right next to the form's outer face about 1 foot away from the house wall (FIG. 7-1). Brace your knee or foot next to the form's inside face while lightly pulling the top of the stake toward you. By sandwiching forms between your foot and the stakes, you are better able to control and guide stakes as you pound them into the ground. Putting stakes into the ground correctly takes practice. They should be straight up and down in order to secure forms in a vertically upright position.

Lay the form against the stake and check it for square. With the form's top edge at the designated mark, or slightly higher, brace your foot against its inside face and drive a duplex nail into the outer face of the stake until it goes through, and into, the form, securing the form to the stake.

Check the square again. If it is off, move the far end of the form horizontally one way or the other until square is established. Once that is done, pound a stake next to the outer face of the form about 1 foot in

7-1 The first stake placed along your side form should be about a foot from the end of the board. Brace your foot or leg against the form to keep it steady.

7-2 Once the form has been positioned, drive a second stake into the ground about a foot in from the end of the form.

from the far end (FIG. 7-2). Use the same foot bracing technique as you did for the first stake. This method is essentially used for placing almost all stakes. Do not nail this stake yet, as the form's proper level must be established first.

Place your level on top of, and in the middle of, the form (FIG. 7-3). Because this side form will run away from the house, the end next to the house should sit higher than the far end. This is designated on the level tool when the bubble is located past the indicator mark closest to the house. Bubbles always rise toward the high side (FIG. 7-4). If necessary, use a square-point shovel to add fill dirt or cut down high spots until the form sits in a slight, downhill run away from the house; all the while making sure the top edge butted next to the house remains at, or slightly higher, than its designated mark.

As a general rule of thumb, outdoor patio slabs, and the like, should fall about 1/4 inch per foot to ensure adequate water runoff capability. If your carpenter's level has two indicator marks on each side of the bubble, plan to position forms so that the bubble touches the second line on the high side. For just single indicator marks, let bubbles rise about an eighth of their length past the mark to establish downhill runs. Once your level indicates appropriate slope, nail the second stake to the form.

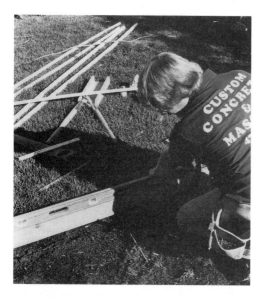

7-3 Before nailing forms to stakes, put a carpenter's level on top of forms and check slope. The bubble inside your carpenter's level will rise toward the high side.

7-4 The bubble goes past the left side mark to show that the form is higher on that side. A good slope is when about one-eighth of the bubble is past its mark on the designated high-point side.

7-5 An easy way to set forms at their precise height is to nail them to placed stakes while they sit too high. Once nailed, lightly tap stakes down until forms reach their appropriate height.

Now, if you allowed the form to ride a little higher than the mark on the house, don't worry. You can always pound stakes down a little to force forms down (FIG. 7-5). The nail securing the form and the stake together allows both to move down in unison. If form ends are too low, however, it is not advisable to pull stakes up. In this case, simply pull nails out of the stakes and reposition forms for new nails.

Your next job is to place stakes along the middle of the form at intervals of no more than 3 feet. To ensure form straightness, place a small nail at each end of the form as close to the inside, top corner as possible; then tie a string to one end. Pull the string tight and secure it to the other end (FIG. 7-6). Positioned along the top, inside edge of the form, this string will serve as a perfectly straight guide (FIG. 7-7). As you position stakes along the outside face of the form, push or pull the form until its top, inside edge is in line with, and runs flush to, the string. Pound stakes into the ground as required. Check and adjust the level before driving nails.

Setting stakes can be tricky work. You might have to hold both the form and stake in one hand to push or pull them into position while at the same time pound on the stake to secure it. In rocky soils, you might need to use steel stakes in order to achieve minimum stake depth for form support. Soft soils require long stakes and short stakes are fine for most hardpan surfaces. Should an end be forced out of square while staking, you can use a kicker stake to force it back where it belongs. Kicker stakes are inserted at an angle from the top edge of forms toward the outside (FIG. 7-8). These stake faces will only ride against the top outside corner of forms to force their top edge in toward slab areas (FIG. 7-9).

Should you be required by the building department to place a footing under the perimeter of your slab, you'll need to use long stakes, which will be difficult to place (FIG. 7-10). Many professionals have discovered it is easier to form jobs first with long stakes and then dig out footings once

7-6 A string is secured to each end of the form to serve as a straightedge. It is located right along the form's top, inside corner as pointed out by the pencil.

7-7 The string guide helps the form installer position stakes in such a way that they keep the form straight. It might have been easier to form this job first and then dig the footing.

7-8 Kickers add support to form tops and help keep them straight. Kickers are extra important on this job because regular stakes have little ground to support them due to the footing.

7-9 Stakes laid across the footing help support the form. A number of kicker stakes are spread out in preparation.

7-10 A form installer braces a long stake at the top with his hand while using his foot to brace the bottom. A string guide is in place and is used as a straightedge.

7-11 If forms must be placed off the ground to establish slab height, a second form can be placed directly under a top form to fill the void.

7-12 Be alert to the guide strings position when placing stakes. Caught on a sliver or other obstacle, it will be off center. Footings are easiest to dig after forms are installed, especially in softer soils. Use a square-point shovel to square footing sides.

all the forms are secured in position. With forms in place, they are able to accurately determine footing perimeters and can dig perfect footings straight down from inside form faces using a square-point shovel (FIG. 7-11). This eliminates the need to bridge forms in place while attempting to insert stakes, a process hindered by the collapse of footing walls (FIG. 7-12).

SECOND FORM

The second form of a slab poured against a house or other structure's wall is the one that runs parallel to the first. One end of it will butt against the house and run slightly downhill away from it. The end form is saved for last. That is the form that parallels the house and designates how far the slab will extend out from it.

This second form will be located a predetermined number of feet away from the first form to provide your slab's length; the distance in which the slab will run along the wall of the house, as opposed to out away from it. The easiest way to pinpoint the location for placement of the second form's near end is to measure across the wall from the end of the first form. Remember, when you attach your tape measure to the outside face of the first form in order to stretch it across the wall, the width of your 2-×-4 form is 1½ inches. Therefore, if the length of your slab is 20 feet, measure 20 feet 1½ inches. The compensating inch and a half allows for the width of the first form. Your mark should then be exactly 20 feet from the inside face of the first form to the inside face of the second form.

At the spot where the second form butts next to the house wall, there should be two marks—one to designate the level (top edge of the form) and one to designate the length of the slab (inside form face). Using your marks as described for the first form, set the first stake, check for square, level, then nail (FIG. 7-13).

7-13 Second side forms must be positioned square with the buildings they come off of and parallel to first side forms. Use a tape measure, carpenter's square, and string to assist in form placement.

You can set the end stake for this second side form in three different ways: with a carpenter's square, as you did for the first form; by using a tape measure stretched from the far end of the first form to a length equal to that along the house wall; or by laying the end form in position, with either an actual pencil mark denoting the proper length (space) between side forms or with it being cut to exactly the correct length. Again, you must check for square and level and adjust the form accordingly. Once you are assured that this second side form is positioned correctly, with stakes at each end, place middle stakes at least every 3 feet (FIG. 7-14).

7-14 Brace your foot against forms when driving nails through stakes and into them. If nails are long enough to penetrate through a form's inside face, be sure your foot is off to the side of the nail's position.

END FORM

As long as both side forms are installed correctly, the end form should be easy to position. If this form is cut to size, you can simply measure down each side form and mark the point where you want the end form to be. On a slab measuring 10×20 feet, for example, you would make pencil marks on the top of both side forms exactly 10 feet from the house wall. The ends of the end form would then be nailed to the far ends of each side form at the prescribed 10-foot pencil marks. As long as both side forms are set to grade (proper slope level), and you nail the end form so that its top side matches the side form's top sides, its level should be correct (FIG. 7-15).

A word of caution; just because lumber is supposed to be in even-

7-15 Nails are driven from the outer face of a side form and into the end of an end form. Hold end forms in position while nailing.

foot increments does not always mean boards are cut exactly to length. A 10-foot-long board, for instance, could easily measure 10 feet, 2 inches, or any such length just over what you would expect. Therefore, it is always a good idea to measure form boards before using them.

With both ends of the end form securely nailed to the side form ends, place stakes as you did for the other two forms. Be sure to check for level as you go (FIG. 7-16). Use a string to ensure the form is staked straight. To double check corners, use a carpenter's square (FIG. 7-17).

In lieu of relying on a string guide to help stake the end form straight, you might be able to use a block. In this case, a block would be a 2-×-4 board that measures the exact distance you want the slab to extend out from the wall; on a 10-×-20-foot slab, it would be a 2×4 that measures exactly 10 feet. Lay this board on its narrow side with one end against the inside face of the end form at the point where you want to insert a stake. Place the other end of the 2-×-4 block against the house wall and check square.

At this point, stand on the outside of the form, and place the stake against the end form with your foot braced against it, essentially wedging the stake between your foot and the form with the block holding the form steady. This should allow you to quickly stake the end form with its inside face exactly 10 feet away from the house wall. Check the level and then nail the stakes to the form. Using blocks to support forms while installing stakes is a great time-saver, especially for patio slabs, narrow jobs, and walkway projects.

7-16 A carpenter's level rests on an end form as a stake is pounded into the ground. Before this form is nailed to the stake, it is checked for proper slope.

7-17 A small carpenter's square is used to check a 90-degree corner while setting a form. The top form has already been secured.

KICKER STAKES

Kicker stakes were mentioned earlier in the chapter as a means of forcing top form edges in toward slab areas when they have been accidentally forced out while installing other stakes (FIG. 7-18). Not only can kickers adjust form edges, they also add a great deal of support to form tops. This

7-18 Kicker stakes are driven in at sharp angles along top, outer form corners.

is especially important for pours with extra long runs where you want to be sure top concrete edges are as straight as possible, such as patio slabs that run the entire length of a house or a long driveway (FIG. 7-19).

Kickers are an absolute must for jobs that require footings. This is because upright stakes might only have an inch or two of dirt between them and the open footing (FIG. 7-20). Because of the enormous weight of concrete, all footings are completely filled with concrete to the base of forms (grade level) first, before any concrete is placed against forms to create an actual slab. This gives footing concrete time to settle and actually lend support to upright stakes by filling the footing void.

GRADING

Once forms are secure, the outline of your concrete job will be clear. At this time, you must concentrate efforts on making the entire area inside forms conform to an even depth of no more than 4 inches. This goal is referred to as *grading*. Endeavors to flatten slab areas to an even depth of $3^1/2$ to 4 inches accomplishes two things: precise dimensions for calculating concrete yardage, and an evenly dispersed concrete depth to help avoid uneven pockets that could lead to eventual cracks. Haphazardly graded slab areas might result in some spaces being filled with 5 to 6 inches of concrete while others only get 2 to 3 inches. Uneven depth causes weak spots that generally crack.

7-19 Plenty of kicker stakes have been placed along this driveway form. The form is solidly secured in position and should not budge during concrete pouring.

7-20 Kickers are required for forms along footing installations to keep forms from "blowing out," or moving out of plumb.

Grading work is most labor-intensive when dirt has to be removed. You might be able to simply scrape soft soil from the surface with a square point shovel, if you're lucky. But more often than not, you'll need to use a wide-blade pick to chop dirt loose, which is then scooped up with a square-point shovel. The grading tips in this section can help minimize the workload of grading and avoid digging more than necessary. These tips help you to set up a grading system that quickly checks excavation depth as you progress across slab areas.

The simplest way to determine grade depth is by placing a screed board on top of forms and measuring the depth below it (FIG. 7-21). The bottom of a screed board determines the top of a concrete slab. Having 3¹/2 to 4 inches open under the bottom screed board edge is great as long as this is a consistent measurement. If you find that the ground you are working on is extra hard, try using the pick end to break up dirt and the wide blade to clean areas out. Do not get carried away and dig deep holes. Rather, rely on short pick strokes applied at an angle to chip dirt out, not hefty blows directed straight down.

7-21 Use a screed board as a guide while grading base material. Periodically check distances between the bottom of the screed and the ground to ensure even depths of 3¹/2 to 4 inches.

When dirt has been broken up and loosened, you might find that a heavy-duty rake works well to move material out of the way and into piles for later pick up with a shovel (FIG. 7-22). Rakes are good leveling tools and can move a lot more dirt than you might think. In addition, your rake might measure close to 3¹/2 to 4 inches and be a perfect tool for grading

7-22 A heavy-duty rake works well for grading bases.

under screed boards to fine-tune depths. Operating a proper-sized rake along screed boards can give you quick and constant depth reference, as every time the rake fits under the screed board clearly, you'll know that the grade depth is close to 4 inches.

Another way to help determine grade inside formed slab areas is by attaching a grade screed to the bottom of your regular screed board (FIG. 7-23). To make this guide, simply nail two wooden stakes to a 2-×-4 grade screed that measures a few inches shorter than your regular screed board. Make the tops of the stakes flush with the bottom of the grade screed so that when it is scraped against the ground, the stakes won't dig in. Then, simply butt the grade screed to the bottom of the regular screed board and nail the stakes to the regular screed.

In essence, you'll have a screed board resting on top of the forms with a 2-×-4 extension attached to the bottom of it and secured by wooden stakes nailed to both of them. This shorter-grade screed should fit between forms, and as long as the regular screed board remains in contact with forms and the grade screed barely scrapes the ground surface, you'll be assured of grading to a 3 1/2- to 4-inch level.

Another way to gauge grade depth is with string. With nails positioned along side forms, you can stretch string across slab widths and periodically measure from the ground up to string locations (FIG. 7-24). In lieu of constantly using a tape measure to gauge grade depth, make a dark pencil line on a stake 4 inches from an end and use it as a measuring

7-23 This upside down stake is nailed to a regular screed board that rests on top of a form. It is also nailed to a shorter, 2- x -4 board directly under the screed. The bottom 2 x 4 is a grade screed that accurately helps to denote proper grade depths.

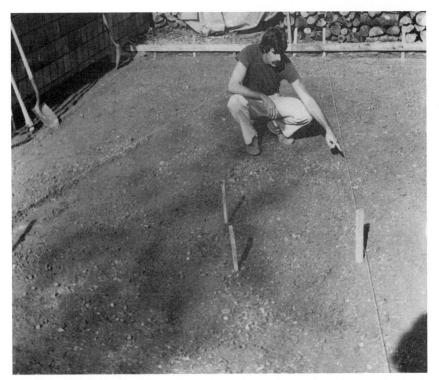

7-24 A string tightly stretched across a formed area shows where the top of a slab is designed to be and assists grading efforts. It was set up with a carpenter's string level and is supported in the middle by a stake to prevent it from bowing.

7-25 A dark pencil mark 4 inches from the end of this piece of wood is used as a quick and handy depth gauge.

device (FIG. 7-25). This is not always the easiest method to use. String can get in the way of excavation work, and it is not nearly as maneuverable as a grade screed.

Sometimes, grade levels are much lower than expected. In this case, you must use fill material. Suppose, for example, you want to pour a patio and you want the top of the concrete to be 2 inches below your sliding glass door sill but the existing grade level is 8 inches below the door. The easiest way to prepare this site and still maintain just a 4-inch slab depth is to form it with 2-×-6 lumber and fill in the area with 2 inches of sand. Optimally, you should leave a 4- to 6-inch-wide trough along the 2-×-6 forms where the concrete will extend from the bottom of forms to the top. This trough will allow concrete to be about 6 inches deep along forms. This way, once the slab is poured and the 2-×-6 forms removed, the concrete sides will extend to the ground instead of exposing a 2-inch base of sand, which could erode.

Sand is an excellent base material for concrete. It grades easily, fills low spots quickly and does not shrink or compress once concrete is poured. In lieu of fill sand alone, you could use pea gravel or a mixture of pea gravel and sand to give your slab a solid and uniform grade base.

SETTING SCREEDS AND SCREED FORMS

Once concrete is on the ground, you must screed it to flatten and level its surface. This phase of any concrete pour is very important. If not done correctly, slab surfaces could be pockmarked with low spots that invite puddles of water and are riddled with waves of high spots that cause benches and tables to wobble, people to trip, and other unnecessary problems. Most small- to medium-sized flat work slabs and walkways enable screed boards to rest directly on top of forms as they are pulled down to level concrete (FIG. 7-26). For larger slabs—longer than 14 feet—

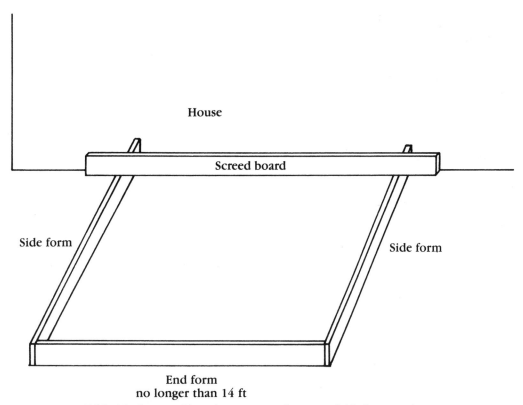

7-26   Ideally, screed boards are operated on top of side forms as shown.

professionals prefer to install screed forms on which one end of a screed board rests while the other end sits on an actual form (FIG. 7-27).

Screed forms are 2-×-4 boards positioned in such a way that screed boards can be operated on them instead of on top of actual form lumber. They are used for concrete jobs where the spacing of actual form boards is too long or too wide to adequately support a screed board. Generally, if this spacing is more than 14 feet, a 2-×-4 screed board will tend to bow in the middle, causing concrete surfaces to incur a slightly rounded indentation that conforms to the bowed, 2-×-4 screed board.

Although you could install screed forms inside slab areas to the same level as side and end forms, it is easier overall to place them above. This is so you will not have to pull them out of wet concrete after screeding and then fill in the void made by their removal.

Placing screed forms above slab areas means that their bottom side will rest on a form's top side. The screed form will be located at the same level the screed board is operated (FIG. 7-28). In order for screed boards to be supported by a screed form in this manner, a wood stake is nailed to the end of the screed board and allowed to stick out 3 to 4 inches (FIG. 7-29). It is this ear that will rest on the screed form while the screed board

7-27 A 2-×-4 screed form end rests on an existing concrete garage apron while supported in the middle and on the other end by screed pins.

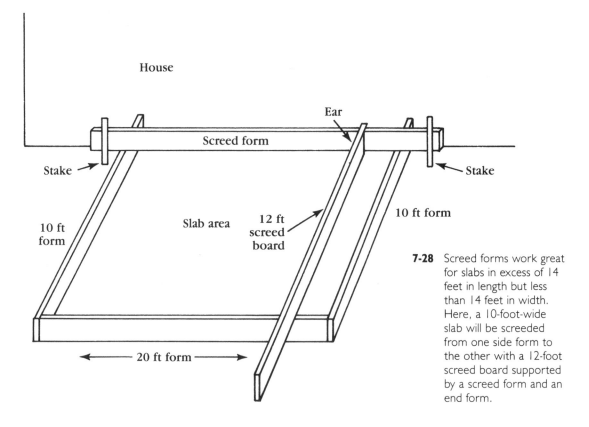

7-28 Screed forms work great for slabs in excess of 14 feet in length but less than 14 feet in width. Here, a 10-foot-wide slab will be screeded from one side form to the other with a 12-foot screed board supported by a screed form and an end form.

is pulled across the slab. The other end of the screed board generally rests on an existing form board. Ears must extend past the end of screed boards by at least 3 inches to allow enough working room for the board to be "see-sawed" back and forth across a slab.

Screed forms are also commonly used on jobs where no regular forms are available to support a screed board (FIG. 7-30). Wide porches frequently present this type of problem. In these situations, screed forms

7-29 To operate screed boards on screed forms, an ear must be secured to an end of the screed board. This system of screeding concrete works well for spans of 14 feet and less. Be alert that screed board ears remain in position during the screed operation.

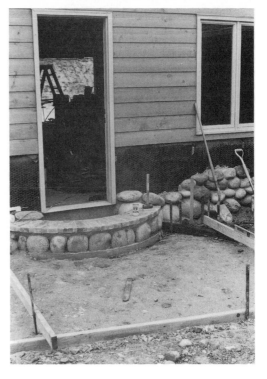

7-30 Screed forms set up on each side of this flagstone and rock porch will support a screed board with ears on both ends of it. Concrete will be placed next to the porch first and gradually filled in as screeding continues toward the form located at the bottom of the picture.

can be installed perpendicular to walls and other obstacles and supported on their open ends by sturdy stakes. When two screed forms are used to support a screed board, ears must be installed on both ends of the screed board.

Screed forms can also be set up to run parallel with an end form for special needs (FIG. 7-31). Each concrete pour is a little different from others and you might have to devise a unique screed form/board combination. Be aware, however, that anything placed inside a slab area that doesn't belong there, will have to be pulled up after concrete is placed. This could become quite a project if you have a number of screed forms buried at any one time. It is much easier to place screed forms above slabs so that the only thing you have to pull out are stakes. Stake holes are quickly filled with partial shovel loads of concrete and quality tamping makes their impressions in wet concrete disappear.

Installing screed forms entails the same basic work you completed on regular forms. The only difference is that the bottom side of any screed

7-31   The long screed form in front facilitates screeding in an area toward the bottom of this picture. The short screed form at the top will be used just for that concrete section located between the two screed forms.

form will determine the concrete's surface level. A carpenter's level is needed for screed form placement. Use it on top of the screed form to help you set the "free" end when situations do not allow the form to span across two regular forms. The free end will be supported by a stake and the other end will rest on a form.

Once a stake has been pounded into the ground, lay a screed form against it about an inch or two too high. Once it is nailed, put a carpenter's level on the screed form. Gently pound the stake down until the screed form reaches the desired level. Screed forms should be supported by stakes at 4-foot intervals to ensure they are not pushed down while operating a screed board on top of them.

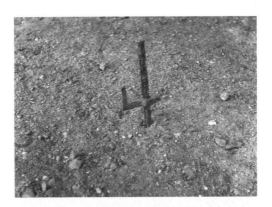

7-32 The arms that extend out from screed pins are adjustable. They have to be in order to meet grade requirements in all soils.

7-33 This screed pin was driven into the ground securely and then its arm adjusted a bit higher than grade as determined by the string guide. With a screed form in place, the pin is hammered down until the bottom of the screed form perfectly matches the string level.

Professionals use screed pins to support screed forms in the middle of slabs (FIG. 7-32). Round steel stakes support metal arms that measure about 1½ inches wide—perfect for holding a 2-×-4 screed form in place. Once screed pins are driven into the ground and a screed form is placed on them, installers can put a level on the form and pound screed pins down until the proper screed form level is achieved. String can also be stretched across a slab area to denote grade. Again, with a screed pin and form positioned high, the pin can be hammered down until the screed form meets flush with the string (FIG. 7-33). Be certain the string is taut; loose string will bow in the middle and sit too low.

By taking a few minutes to visualize your concrete pour in action, you should be able to determine from which direction concrete will come and how it can be properly screeded. Remember that concrete is heavy and screeding is a labor-intensive chore. Long screed boards will have to move and flatten a lot of concrete. Although you should be able to screed spans up to 14 feet with just one screed board, three workers will be needed—two for the ends and one in the middle. If a screed form was set up in the middle of such a pour, you could screed half the area with a shorter screed board that only requires two operators and much less physical effort.

Chapter 8

Forming walkways

*C*oncrete walkways are formed with 2×4 lumber, much the same as larger slabs. About the only difference is the drastic change in design dimensions. Walkways are generally long and narrow. Although forming slabs featuring narrow widths is relatively easy, maintaining perfectly straight sides along the entire length of long walkways can be troublesome. Be prepared to use string as a guide or plan to use blocks while forming. Walkway forming work requires an adequate amount of time be spent on overall design and installation planning. Will you want concrete poured right next to your house, or will a planter between the house and walkway be more visually attractive? Will you need to install water or electrical conduit under the walkway for a future landscape sprinkler or lighting system? How about rain gutter downspouts? Will they require drains? And, consider the walkway's width. Most of the larger concrete tools you'll use will be 3 feet long, like the tamp, bull float, and fresno. Will finishing work be made easier if the walkway was formed at 3 feet, 1 inch instead of exactly 3 feet or less?

About the easiest way to ensure board straightness while staking and securing walkway forms is to use blocks. These are 2-×-4 boards cut to the precise width of your walkway (3 feet, 1 inch) and inserted between a featured wall and the forms (FIG. 8-1). With one end of a block butted next to the house or other wall structure, you can lay forms against the block's other end to guarantee that the form is placed exactly the right distance from the wall as planned. Stakes are also placed next to forms in line with blocks. With your foot braced against a stake, sandwiching a form between it and the block, you can pound the stake straight into the ground easily and at the same time keep the form correctly positioned.

Because the same block, or a set of matched blocks, are used throughout a walkway forming job, forms will always be positioned away from walls at an equal distance and also be straight. About the only problem encountered with this system is uneven structure walls, such as those

8-1 A 2-x-4 block is braced against a wall behind the form installer and is used as a brace for the 2-x-4 form in front of it. Blocks help form placing and staking jobs go faster because predetermined width dimensions are continually maintained by rigid and unison block support.

covered with plaster or stucco. On these, you will have to use a string guide.

If your walkway is designed with a planter between it and your house, you can still use blocks to aid forming. Simply cut blocks equal to the width of the planter and use them to install the inside form—the one closest to the wall. Then use longer blocks cut to the walkway width to place the outside form. Just butt them up to the staked and secured inside form.

LAYOUT

Walkways are laid out the same as slabs. Forms are placed in their approximate locations to make best use of each board. Use forms to their fullest potential and try not to cut any unless absolutely necessary. Lumber is expensive and many concrete forms can be put to good use after concrete jobs. Completely cleaned with water and a scrub brush, a lot of do-it-yourselfers have used form lumber to build patio covers, sheds, and other outdoor structures.

Many professionals prefer to string walkway jobs before pounding stakes and nailing forms (FIG. 8-2). Stakes are placed in strategic positions at the far ends of a proposed walkway to line up a guide where forms are to be installed. String levels help to accurately determine where to secure

8-2 The walkway/porch area in the front of this house opens in the middle to extend out into another perpendicular walkway to the right. To maintain identical in-line form placement on both sides of the open area, a string is stretched and used as a straightedge.

string on stakes to designate slope. Tightly stretched string serves as a straight-edged guide for forms as well as a slope indicator.

Once all of the height, width, and square adjustments are made with string, form installers can quickly place forms, drive in stakes, and nail the package together accurately. Always be alert to string that hangs up on form slivers, stakes, or nails when working with string guides. Likewise, always be certain string is stretched tight to eliminate bows.

SLOPE

To allow for adequate water runoff, walkways must also be formed and poured with a definite slope running in the direction water should drain. This is a very important consideration. If concrete is poured in such a way that water is able to puddle next to structure walls, moisture can eventually "wick" its way into them, causing dry rot, mildew, and other water problems. At the least, always be sure walkways fall away from walls. This is accomplished by having form ends raised higher along walls than their free ends, which extend out.

Typically, walkways are designed to connect patio slabs in a backyard with driveways or porches in a front yard or just run from the backyard to the front. In most cases, front yards and backyards feature different grade levels. In order for the top of a walkway to meet flush with both a rear patio and a front driveway, it might have to run downhill. This is perfectly fine. However, even though this walkway will slope from front to back, or vice versa, you still need to make certain it also falls away from the house to ensure adequate water runoff.

If your patio and driveway are situated in such a way that both surfaces sit at the same grade level, you can pour a connecting walkway that runs level from front to back yards; as long as slope away from the house is maintained.

The rule of thumb for slope on most outdoor concrete flat work projects is a pitch of 1/4-inch vertical drop per foot of horizontal run. Extending from a house wall, for example, a 3-foot-wide walkway should have a drop of at least 3/4 inch; with the farthest edge from the house wall being 3/4 inch lower than the edge against, or closest to, the wall. This drop must be maintained the entire length of the walkway.

The easiest way to guarantee proper slope is to use a carpenter's level. First, make a pencil mark on the back corner of the house wall to designate where you want the top of concrete to be at that point. Do the same at the front house wall corner. Next, snap a chalk line to connect both pencil marks. Because forms will be laid out and stakes anchored into the ground after using the block system, you can rest one end of a screed board on top of the loose form with the other end matched to the chalk line. Put your carpenter's level on top of the screed board and move its far end up or down until it is level.

Once level, you can measure down from the far end 3/4 inch, 1/4 inch per foot of run on a 3-foot-wide walkway. Adjust the form so its top meets with that 3/4-inch drop designation. It should touch the bottom edge of the screed board. Prop the form up with dirt or rocks to hold it steady. Now, take note of the actual location of the bubble on your carpenter's level with regard to its adjacent line marks. Is the bubble about an eighth of its length past the line closest to the house? At the least, it should read as definitely being "off-level" with the high side toward the house. If half or more of the bubble length is past its line mark, the slope is too great. Anything more than a 1/4-bubble length past the mark is too much.

After determining where the bubble on your level will rest while denoting a 1/4-inch slope, you'll be able to lay screed boards on top of forms with the other end flush against the chalk line on the house. Adjust forms to proper height by simply reading the level. You will not have to measure a 3/4-inch drop each time. Do this at the location of each stake so you can nail forms immediately while still keeping your level in sight.

SCREEDS

Unlike a lot of concrete slabs, most walkways are not formed with boards on all four sides (FIG. 8-3). Generally, a house or other structure wall serves as one side. Screeding such walkways can be accomplished in two ways.

Professional concrete finishers frequently wet screed walkways that are poured against walls. They snap a chalk line on walls to designate where the top of concrete should be. Because that line will probably be

8-3 This front porch area requires extensive use of screed forms as a systematic means to accurately screed concrete once it is on the ground. You might have to develop an equally innovative system to achieve acceptable pouring, screeding, and finishing results for your own job.

covered with mud as the pour progresses, they place a second chalk line exactly 3¹/₂ inches above it.

Because 2-×-4 boards actually measure 3¹/₂ inches wide, the distance between these two chalk lines will be equal to a 2-×-4 screed board. Once concrete is on the ground and hand floated up to the bottom chalk line, a screed board is operated across concrete with its top edge flush with the top chalk line. In essence, the bottom chalk line will designate the correct position for the bottom edge of the screed board and the top chalk line will serve as a guide for the top edge. A keen eye is required for this type of free hand work. Not only must you keep the far end of the screed board on top of the form, but you must be alert to low spots, high spots, and obstructions, all the while maintaining the board's edge on its designated top chalk line.

A more secure way of screeding walkways is by installing screed forms (FIG. 8-4). You can use screed pins or stakes. For extra long walkways, screed pins might be the best option because you can simply slide one long screed form down to the next set of pins after screeding one section of the walk. This eliminates the need to purchase screed forms for the entire length of your walkway. Note: screed pins must be positioned close enough together to fully support a screed form as it is moved from section to section.

Set up screed forms using string guides or chalk lines, whichever is easiest and most versatile. Don't forget that you will have to put an ear on top of your screed board to ride on top of the screed form.

8-4 Screeding maneuvers in this area are best accomplished from the house out to forms. Each screed form support stake is pulled after concrete is poured and screeded, which can be done while tamping.

EXPANSION JOINTS

A combination of expansion joints and control joints (seams) should adorn just about every walkway more than 6 feet long and bordered on either side with another concrete slab. As a rule of thumb, expansion joints should be placed every 10 to 12 feet, divided evenly by the insertion of seams.

A 24-foot walkway, for example, should have an expansion joint in the middle and at each end that is poured against any other concrete slab. Because the middle, in this case, is at 12 feet, seams could be spaced 3 or 4 feet apart to offer a visually symmetrical pattern of both expansion and control joints.

Seams, as you will recall, are grooves made in concrete (FIG. 8-5). For all intents and purposes, they are designed weak spots where concrete should crack first, if it is going to crack at all. A crack along a seam is not nearly as noticeable or unattractive as a crack jogging awkwardly across a concrete surface at various angles. Seams are created by special tools called seamers. As a guide for the seamer, 2-×-4 boards are laid across walkways. Nails driven partially into outer form faces are used to designate predetermined seam locations.

Expansion joint felt is a material specifically made for concrete expansion joints. It is used on city sidewalks, driveways, and curbs. Also, most cities and counties require felt be placed next to public sidewalks, driveways, and curbs whenever new concrete is poured against them.

8-5 The deep groove on this walkway is referred to as either a seam or a control joint. It is a weak spot designed to control concrete cracks.

Because the material is not rigid, it has to be supported inside forms with a 1-×-4 brace. That brace is held in place with stakes. Felt braces and stakes must be positioned on the far side of felt so that concrete can be poured next to felt first and then against the brace. This allows concrete to hold felt next to the brace and stakes while mud is poured behind them. Doing it the other way around would cause the brace and felt to be pushed away from stakes.

Once you have placed concrete against felt and about a foot past the brace, you can remove the stakes. Use a hand float to move mud into the holes made by removing the stakes. Then, while slowly pulling up the 1-×-4 brace, force mud under it with a hand float to maintain pressure against felt. Add concrete as needed to fill in voids left by brace removal. If you do it right, plenty of concrete should be positioned against the felt's backside so that it remains in an upright position undisturbed, secure, and straight.

Stringers

In lieu of felt expansion joints, many concrete finishers prefer to insert pressure-treated, 2-×-4 stringers in walkways (FIG. 8-6). Placed inside forms, their ends can be nailed to side forms for easy installation. For walkways poured against walls, one stake is usually sufficient at the wall end while a couple of large nails will hold up to 3-foot-long stringers securely against forms and keep them straight. As with a felt expansion joint, the lone stake is pulled after concrete has surrounded the board.

When using pressure-treated 2×4s as stringers, you must drive nails partly into both side faces. Protruding nails encased in concrete serve as anchors for stringers to prevent them from floating out of concrete during

8-6   All walkway corners must feature either control joints or a combination of control and expansion joints to prevent cracks.

wet weather and from lifting out during cold weather when concrete contracts.

Stringers can be placed anywhere along walkways. Rather than spaced 10 or 12 feet apart just for expansion joints, you can install them every 3 or 4 feet to cover both expansion and control joint needs. The design is up to you and your creative imagination.

Regular 2-×-4 lumber cannot be subjected to outdoor weather year after year. All too soon, it will suffer dry rot and simply break apart. Therefore, always use pressure-treated lumber for stringers. Some finishers have experienced good luck using redwood and cedar, but most prefer material that has been specially treated to stand up under harsh weather conditions. If you are not satisfied with the plain appearance of pressure treated lumber, plan to stain the tops of stringers after concrete has cured. You can stain them a color to match your house, patio, etc.

Benderboard expansion joints

Another alternative to felt material for expansion joints on most walkways is redwood benderboard (FIG. 8-7). Benderboard normally comes in lengths of 10 feet, 4 inches wide and about 3/8 inch thick. You can install it the same as felt (it is not rigid and must be braced) or split it into narrow, 2-inch strips and put them in wet concrete right after it's tamped and bull floated.

The advantage of using 4-inch benderboard instead of felt is that it is easier to operate a hand edger against. The advantage of using 2-inch benderboard strips is that they do not have to be braced, can be simply inserted into wet concrete, and are easy to hand edge against. But, because of its thinness, 2-inch strips of benderboard any longer than 3 feet are very difficult to keep straight while placing them into wet concrete.

Wet concrete will quickly cover pencil marks, so plan to use nails

8-7 A piece of redwood benderboard is used as an expansion joint on one side of this walkway corner. Notice the perpendicular seam that squares off the corner. This setup should prevent cracks from occurring inside the walkway corner.

driven into form sides to designate the location of benderboard strip expansion joints. Strips that are cut an inch or so shorter than walkway widths allow a little extra maneuvering room for wriggling them in place while keeping their entire length straight. If need be, gentle hammer taps can be used to persuade strips into position. Lay a board on top of strips and smack that board with a hammer to prevent denting or splitting benderboard. Afterward, use a hand float to smooth concrete around strips.

Should a section of benderboard stick up after concrete has cured, cut it flush to the walkway surface with a razor knife. Sometimes, you can run the blade of a shovel along protruding benderboard strips to split it and make it flush with concrete.

Chapter **9**

Concrete pouring and placement

*B*y far the most physically demanding and time-constraining parts of concrete work is the actual pouring, placing, and screeding of concrete. Although the only way to avoid such labor-intensive work is to hire someone else to do it, there are ways you can do it efficiently, economically, and with a minimal amount of effort. Above all, don't fret time constraints with regard to concrete delivery truck overtime fees. Hurrying a pour without regard to the disposition of screeding maneuvers can cause a tremendous amount of extra work after the truck leaves. And, the amount of money saved will likely be minuscule compared to the efforts of moving concrete around to accomplish adequate screeding.

THE CONCRETE COMPANY

Large, ready-mix concrete companies generally cater to busy contractors who pour lots of concrete every week. Their schedules are tight with little room to compensate for extraordinary amounts of standby time. Because most of their customers are professional concrete finishers, they can rely on jobs being ready on time and poured within allotted time frames. This is why many large concrete companies might only allow 4 to 5 minutes per yard unloading times.

On the other hand, quite a few smaller concrete companies recognize the inexperience of do-it-yourself concrete finishers and offer unloading time frames up to 10 minutes per yard before any standby fees are incurred. Although they service contractors on a regular basis, their schedules allow for longer unloading periods to better serve smaller residential pours. Likewise, their drivers are more apt to offer advice and recommendations to do-it-yourselfers because they have been involved with so many patio, walkway, and driveway jobs.

The best way to determine which concrete company can meet your needs most economically is to call a number of them on the telephone. Ask about all of their charges and unloading times. Ask the dispatcher if his company makes a lot of deliveries to homeowners or if it mainly serves large contractors. Novice do-it-yourself concrete finishers are recommended to go with concrete companies most familiar with homeowner needs and who have the longest unloading times.

When the concrete delivery truck arrives at your job site, take a few minutes to introduce yourself to the driver. Show him your job layout and explain how you plan to conduct the pour. Mention that you do not have a lot of experience pouring concrete and would certainly appreciate any helpful tips or advice.

Drivers generally have a good idea of what forms, ramps, and other related attributes should look like when properly prepared because they deliver concrete to such a wide range of amateur and professional concrete finishers. They might suggest you install a few extra stakes here and there, provide more solid support for wheelbarrow ramps or may even see a much more efficient and less labor-intensive way to get concrete from their truck to inside your forms. The more congenial you are, the better your chances are of obtaining some very useful advice.

Ready-mix drivers control concrete flow and operate the chute up or down from the back of their trucks. If yours is a job where wheelbarrows are used to deliver concrete to forms, the truck will remain parked and the driver will stand at the back of it to operate controls.

For jobs where concrete will be poured directly from the truck to the forms, drivers will remain in their cabs. From there, they will maneuver the truck and simultaneously operate the chute and control concrete flow according to directions given by a designated person, usually the one maneuvering the chute.

Because the driver must sit in the cab, his vision is limited to side mirrors. Therefore, someone must be within mirror range to provide directions. You could yell and hope the driver hears you, but chances are, the truck's engine noise will limit that means of communication. This is why most professionals use a system of simple hand signals to direct drivers. Be absolutely sure you and your driver have a clear set of hand signals worked out before the pour starts.

Normally, a clenched fist held in the air means to stop the truck from moving forward. This is done when the end of the chute is about 1 1/2 to 2 feet ahead of the last concrete poured. At that point, the chute is swung to the furthest part of the slab and the flow of concrete is started.

If the driver is able to see the job from a side mirror, he will pull the truck ahead as you finish filling a section with concrete. If the job is out of view, you will need to wave an open hand under your chin to designate that you want the flow of concrete shut off. Do this a couple of feet before sections are filled, as concrete will continue to flow down the chute even though the truck's container drum has stopped moving. Extra concrete poured into filled areas will have to be moved out with shovels, a lot of heavy work.

The chute is raised and lowered by a hydraulic ram. During a pour, keep the end of the chute about 1 to 1¹/₂ feet off of the ground. This minimizes concrete splatter and helps the chute operator better judge how much concrete is needed to fill certain areas. To have a driver adjust the chute, point to it and then simply point up or down.

Any set of signals can be quickly worked out with just a little discussion. Keep it simple. Should problems arise, stop everything and verbally communicate. Work together to help your job flow smoothly.

FINAL CHECK

Before the concrete truck arrives, walk the job site to make sure everything is ready to go. Put a little pressure on forms with your foot to make sure they are firmly braced and supported. See that conduit pipe, cleanouts, or other obstacles are either covered or secured (FIG. 9-1). Check the location and condition of tools to have them ready at hand but not in the way.

9-1 Make sure forms are secure and other attributes are correctly braced or supported before concrete is delivered. The white pipe sticking out of the black electrical conduit in this formed area is simply taped in place to hold wires out of the way and prevent concrete from falling into the conduit. Note the hog wire on the ground.

Patio slabs and walkways seldom need reinforcing wire or rebar because they won't have to support heavy loads. Driveways are a different story. Because they will have to support the weight of automobiles and delivery trucks, you might consider laying down a layer of 6-×-6-×-10 reinforcing wire, commonly referred to as hog wire. This material features #10 gauge wire fabricated into 6-×-6-inch squares. It is available through concrete companies, lumberyards, and other building supply outlets. Rolls are 7 feet wide by 50 to 100 feet long. You buy it by the linear foot, which will actually be 1 linear foot 7 feet wide to equal 7 square feet.

Although a layer of hog wire strengthens slabs, it might not do a lot to prevent cracks. Its main function is to keep cracks from separating horizontally or vertically.

Before the actual pour, be sure hog wire lays flat. No part of it should be allowed to stick up past the tops of forms. Also, notify all your helpers that they will have to occasionally reach down with a rake or their hand to pull wire off the ground and into the middle of concrete. Laid flat on the dirt, wire will not do much good. Ideally, it should rest in the middle of concrete and be completely encased.

Your final check must also include the ground base. Is it wet or is it bone dry? During hot, dry weather, base material can almost never be too wet. In extreme desertlike conditions with parched ground and temperatures hovering around 100 degrees, professionals have soaked bases to the point where concrete was poured over water puddles and they still had to work fast to finish it before it quickly dried out.

In cases such as this, you might be wise to set up and flow a lawn sprinkler over bases the night before a pour to ensure that your concrete does not "get away from you" or you "lose it"—terms used by professionals to describe slabs that dried much too fast and did not allow enough time to effect a smooth finish.

Finally, be sure to talk over the job with helpers. Let them know what to expect. If your job is such that the driver will dump half the load and then have to reposition his truck, alert helpers so they can assist the driver by guiding him in and out of tight spaces. Remember, the truck will make a lot of noise and normal voice communications will be hampered. Be especially careful when directing the driver so he does not run over forms, stakes, or helpers.

CONCRETE LOAD CHECK

After the driver has walked the job site and positioned the truck, he will get out and ask if you want to see a sample of the concrete. If he doesn't, you insist. Always check concrete before actually pouring. This gives you the opportunity to have water added to dry mixes. Dry concrete is very difficult to work with. Unless your job consists of stairs or a very steep driveway, wetter concrete is recommended.

Concrete delivery trucks are equipped with huge drums that continuously rotate to keep concrete thoroughly mixed. Welded to their inner drum surfaces are a series of spiral shelves. As long as the drum rotates against these shelves, concrete is forced toward the bottom of the drum. To get concrete to exit drums and flow down chutes, drivers simply reverse the drum's rotating direction. This causes drum shelves to force up concrete material much like a drill bit forces out curly wood slivers.

To see a small sample of concrete, the driver will reverse the drum's direction until concrete flows 2 to 3 feet down the chute. Then he will stop. Look at the mud and determine its consistency. Dry concrete will not flow down the chute easily and will tend to batch up on itself. Wet concrete will flow like water and not batch up at all. Unfortunately, if your load is wetter than you prefer, there is little you can do about it. If the mud is dry, however, you can have the driver add water to the mix. Concrete trucks carry their own water supply for this purpose.

Confer with the driver about the concrete's consistency. Make sure he understands you are a novice finisher and would prefer a wet load to a dry one. If concrete piles up on the chute and seems to have a tough time flowing, have the driver add a half gallon of water per yard of concrete. Once this is done, ask to see another sample.

Good concrete mixes flow easily down chutes. They are also much easier to screed, tamp, and finish (FIG. 9-2). When you believe the mix is just right, allow some to dump on the ground (inside forms, of course). A good mix should not pile up, rather, it should flow outward and essentially flatten itself out. In addition, good mixes will fall off the ends of chutes in a constant flow, where dry mixes tend to plop on the ground in stuttered batches.

It is not uncommon for concrete to start to dry out in the middle of a pour, especially on hot days. If this happens, stop the driver and ask that more water be added to the mix. A half gallon of water per yard of concrete is a good starting point, although a lot of finishers are confident that one gallon per yard is not excessive. Just remember, though, you can always have water added but you can never take water out.

DIRECT POUR

Pouring concrete directly from a truck's chute requires at least three, preferably four, workers. One to handle the chute and direct the driver, two to screed, and one to assist screed operations by moving concrete with a rake or shovel.

The chute operator is responsible for directing the chute in such a way that concrete fills all reachable areas with each pass. Essentially, the chute is moved from left to right on one pass and then right to left on the next, and so on. Each pass should result in just enough poured concrete to fill a swath up to the tops of forms and the bottom of the screed board. Too much will require removal by shovel or rake, a heavy work load. Too little might require another quick pass of the chute or a few shovelsful to fill in. Long, steady sweeps of the chute work best.

The chute operator needs a keen eye for estimating how much concrete falls on the ground. He must also signal the driver when it is time to

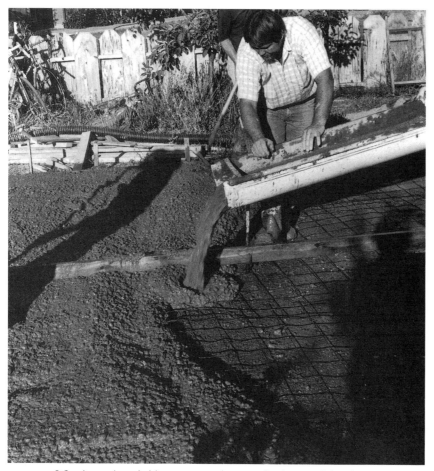

9-2   A good workable concrete mix pours freely out of the chute.

move forward or stop the concrete flow. When drivers have a good view of pours, they can tell when to move ahead and when to stop. Should the job of pouring get ahead of screed workers by more than 4 to 6 feet, stop the concrete and let them catch up. This is really important if too much concrete has been poured, as all of the workers will need to get shovels and rakes to help move excess concrete back to empty form areas. Operating a chute to result in a perfect 4-inch-thick pour takes a lot of practice. Have patience and never be afraid to stop the truck if the pour starts to progress too fast.

Although one worker will be assisting screed operations by moving excess concrete away from the screed board and filling in low spots as they occur, a chute handler can sometimes help flatten and place concrete by pushing it with his foot (FIG. 9-3). This works great for forcing concrete into corners and flattening inadvertent high spots.

If only three workers are available for your pour, the two screeders

9-3 A chute operator can help spread concrete by pushing it around with his foot. Two finishers will screed one side of the driveway after concrete is poured and roughly leveled out with rakes and/or shovels.

will have to work together with shovels and/or rakes to help initially place concrete (FIG. 9-4). Before enough concrete is on the ground to screed, they can do their best to get a smooth, 4-inch-thick slab in place. They should use a rake to push and pull concrete into a flat and even layer, the same as they would with dirt in a flower bed. When about 4 feet of mud is on the ground, stop the flow of concrete. Two workers should operate the screed board, and the chute operator assist them by pulling excess concrete away or filling in low spots with shovelsful of mud (FIG. 9-5).

Working together as a team is crucial. The screeders, chute operator, and extra helpers must communicate with each other for any job to go smoothly (FIG. 9-6). As much as possible, assist each other so that just the right amount of concrete gets on the ground to eliminate unnecessary shovel and rake work. And remember, when the job of pouring starts to get ahead of screeding, stop the concrete to catch up with screed work (FIG. 9-7). This time taken to catch up will help as much as anything else to make your job progress smoothly and efficiently.

WHEELBARROW MANEUVERS

Moving concrete from a delivery truck's location to inside forms is hard work. Concrete is heavy and it takes a little time to get used to "wheeling" it. Have the driver park his truck as close to the job site as possible to make the wheelbarrow run as short as can be expected (FIG. 9-8).

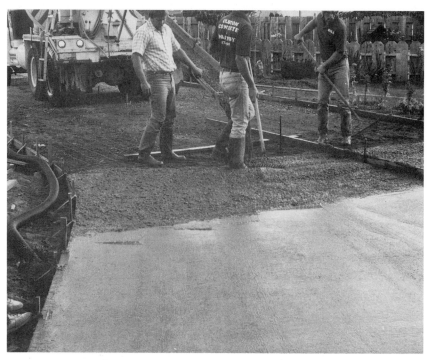

9-4 Teamwork is the key with any concrete flat work job.

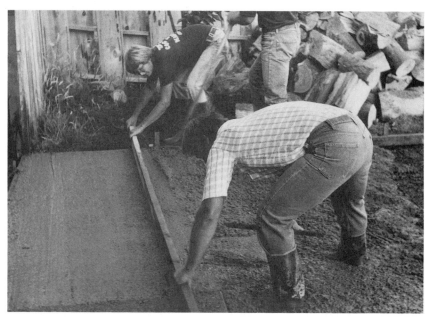

9-5 Screed boards are easiest to operate with two people. Boards are moved back and forth over slabs sideways while being pulled at the same time. Note the worker in the middle removing excess concrete with a rake.

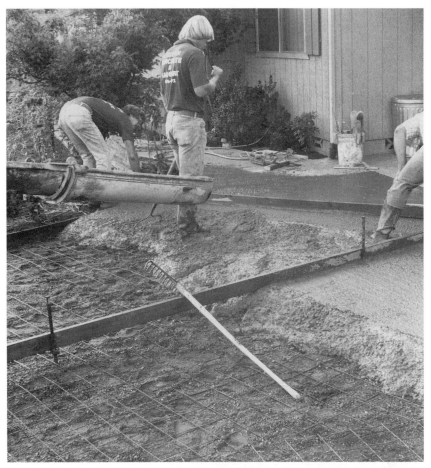

9-6 The screeder on the right is at a screed pin and cannot continue screeding without lifting the board. To ensure a flat screed, the worker on the left has continued screeding. After going about 2 to 3 feet ahead, the screeder on the left will pick up the board and move it back to the spot across from the screeder on the right. At the same time, the board is raised over the screed pin so the job can continue.

At first, have wheelbarrows filled only half full. This lighter load allows users to get a feel for moving the wheelbarrow and its shifting load. It also gives them a chance to determine the best route of travel and what it feels like to go around corners, over ramps, and through tight spots. As the job progresses and "wheelers" get used to the work, they can have the driver pour fuller loads.

If a wheelbarrow should get off course, fall over, and lose its load, all is not lost. Use a shovel to scoop mud back into the wheelbarrow or into formed areas when close by. As time permits, the concrete residue left behind by the lost load can be washed away with water. In worst cases, a wire brush works great for removing dry concrete.

9-7 This screed board was easily maneuvered on top of the concrete apron and form for the section on the left. In the middle of the driveway, it is supported on a screed form by an attached ear.

9-8 Avoid running wheelbarrows over forms. Here, a wheelbarrow has been placed inside the formed area to deliver concrete next to the house where chutes will not reach.

Dumping concrete out of wheelbarrows is done in a smooth rapid movement, much like trying to throw concrete out of it. Many novice wheelers slowly tilt wheelbarrows allowing concrete to gently plop out. Along with making a mess from plopping concrete splatter, slow-motion dumping causes concrete to pile up which, in turn, requires shovel or rake work to smooth out.

By rapidly tilting and throwing up the handles of a wheelbarrow, concrete is forced out and away to flatten and cover a wider area. In fact, done properly, the lip of a wheelbarrow will just about hit the ground first before any concrete is emptied.

Start to tip wheelbarrows about 2 to 3 feet before you reach the edge of concrete already in place on the ground. This way, concrete rapidly forced out of your wheelbarrow will reach up to, and join, that mud just in front of it instead of piling up on top of, or too far back from, the fresh concrete line.

Wheelbarrows always function best when their tire is filled with air to its proper level. Soft tires make wheeling extra difficult. This is especially important when wheeling requires going over ramps, bumps, and around corners. Ramps can be made to go over just about any reasonably sized obstacle. Plywood works good as long as 2-×-4 blocks are placed underneath for support. Going over forms requires that ramps be on both sides to keep from forcing forms out of position. Should you have to go over forms in order to place concrete, consider making ramps at two separate locations. This way, wheelers can get into formed areas from one ramp and exit the area from another to stay out of the way of oncoming wheelbarrow loads.

Because concrete is so heavy, it is best to let your arms hang down while holding up wheelbarrow handles and moving. Relying just on the strength of your bent arms for maneuvering creates a high center of gravity. By allowing your hands and arms to hang as low as possible, a great deal of the power and strength needed to maneuver wheelbarrows is shifted to your legs. With feet spread slightly apart, you'll have much better control of wheelbarrows.

SCREEDING

Screeding, also referred to as rodding off, can be accomplished in basically three different ways: with a screed board on top of forms at both ends; with the use of screed forms; or, by wet screed maneuvers. Screeding is basically just using a straight board to scrape off excess concrete from slab surfaces to flatten them as much as possible.

Walkways not much wider than 4 feet can generally be screeded by one person as long as the screed board can rest on forms at both ends (FIG. 9-9). For slabs wider than 4 feet, plan on having two people maneuver the screed and help pull back excess concrete, as well as look for low spots (FIG. 9-10).

Two people working together on a screed board that rests on forms can slide the screed board back and forth in a sawinglike motion to push

9-9 One person can generally screed areas up to 4 feet wide provided the screed board is supported on both sides. Use a hand float to move excess concrete out of the way or to scoop up small amounts for filling in low spots.

9-10 Constantly be on the lookout for low spots while screeding.

down rock and pull back excess concrete in one step. Simply pulling a screed board straight back tends to make rock tips stick up, leaving small holes behind. A helper behind the screed board and equipped with a rake or shovel, can greatly assist screed operations by alertly pulling back excess concrete or shoving needed mud into low spots (FIG. 9-11).

While screeding, be sure to keep the screed board end in contact with support forms at all times. Once in a while, rocks will land on top of forms and cause screed boards to ride on top of them instead of form tops. Operating a screed board higher than planned will cause high spots

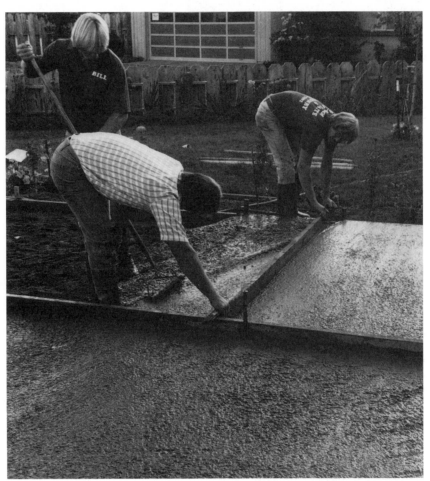

9-11 Use a rake to pull back excess concrete. A slight angled ridge is apparent from the rake toward the far end of the screed board. This is a sign that the screeder on the left stopped at a stake while the right-side screeder continued on. After moving ahead a couple of feet, the right-side screeder stopped and picked up the board to reposition directly across from his partner while the board was raised over the stake.

along slab surfaces. This makes for an uneven screed and a surface that is more difficult to finish because of its uneven nature.

Because you must stand in concrete while screeding larger slabs (narrow walkways can be screeded from outside forms), footprints are left behind every time you move. The easiest way to remedy this problem is to reach behind with your foot and drag some mud to fill holes (FIG. 9-12). When the holes are filled, tap them lightly with the bottom of your foot to flatten them out. Then, screed over them.

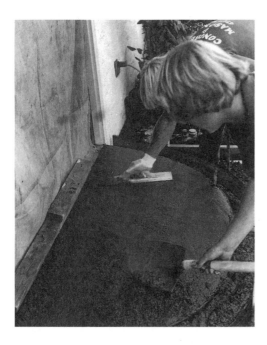

9-12 Fill shallow low spots with handfuls of concrete or small scoops with a hand float. Larger areas can be filled with shovels full. Be sure to screed over all areas that receive extra concrete from any source.

Excess concrete accumulated along a screed board is quite obvious. Low spots are just as important to take care of but might be a little harder to see. Low spots are pockets of concrete that are not filled high enough to reach the bottom of screed boards. Unless they are filled and screeded flat, they will cause slab finishes to be uneven and also serve as spots for water to puddle and dirt to accumulate once the slab hardens.

Low spots are places that look like they have not been touched by a screed board and, in fact, they haven't. Normally, screeding makes concrete surfaces appear smooth and slick because cream is forced to the surface while aggregate is forced down. Conversely, low spots are rough with aggregate and plainly visible. Fill them with a shovelful of concrete or push some mud into their area with a rake, provided they are located close to the screed board. Once filled, run the screed over them to ensure flatness (FIG. 9-13).

It is not uncommon to have to screed surfaces more than once. After screeding a 3 to 4-foot section, check your work to see if any high or low spots linger. If so, go back to the area and screed again (FIG. 9-14). Most of

9-13 Pockmarks in front of the screed board are signs of low spots. After the screed board has completed this pass, workers will fill the spots with handfuls of concrete.

9-14 Use a hand float or your hand to remove excess concrete when screeding over stakes.

the time, second and third screedings are needed for pours where too much concrete was placed to begin with. As the screed board was maneuvered over the area, it was forced to ride on top of concrete because enough strength was not available to move the amount of excess concrete. This is clearly one of those situations where concrete flow should be stopped and all hands used to move excess concrete away from the screed area. Concrete should not flow again until the entire area is screeded properly.

When stakes stick up above form tops, screed boards cannot smoothly pass by. The best way to screed around them is to have the opposite end of the board screed go past the obstruction at an angle. Once the other side has passed the stake a couple of feet, lift the screed board over the stake and have the person handling the opposite side move just ahead of the point where you stopped (FIG. 9-15).

9-15 When obstacles prevent normal screeding, use a hand float to flatten and level out concrete. Here, concrete is carefully placed around a hot tub railing.

By having the opposite end of the screed board go past the stake, you are guaranteed that the area was actually screeded except for a very narrow triangle out around the stake. This small area can then be easily flattened and smoothed through tamping and bull floating.

Should large obstructions prevent normal screeding, try using a smaller screed board. In very small spaces, you can use a hand float to screed. This will require a keen eye for detail and level. Each circumstance is different. Use your ingenuity to solve special screed problems. Don't forget that chalk lines can be useful, especially when placed above the level of concrete so they are not covered during the pour.

Operating a screed board along screed forms presents some extra precautions. Be aware of the screed board ear at all times. Should it slip off of the screed form, it will cause a long divot in the concrete surface. Mud forced out from the divot will be higher than it is supposed to be and will require the area to be completely re-screeded. Another precaution involves screed forms that are suspended over slabs and only supported every 4 feet or so by stakes or screed pins. Because the screed form is suspended, excessive downward pressure on the screed board could cause the form to bow. This allows the bottom of the screed board to fall below expected slab surfaces and causes low spots. To avoid this, concentrate on pulling screed boards back only and not applying any unnecessary downward pressure.

Suspended screed forms should not be buried in concrete. In fact, the only part of them that remains close to the surface is their bottom edge. Therefore, only stakes or screed pins should have to be pulled from concrete and only their holes left behind should need to be filled. Because these holes will be rather small, you can wait to pull them until operating a tamp close by. There is no compelling reason to walk out on fresh concrete just to retrieve screed forms and stakes when you will have to walk out there to tamp soon anyway. Besides, you can easily cover and flatten your footprints with a tamp.

A tamp operator will need help getting screed forms and stakes out of the slab area. He can loosen stakes with a hammer and then pull forms with stakes attached all in one action. Helpers will have to stand by next to forms to get the board from him. If the boards are too long, he might have to remove nails, hand the board to you or a helper, and then retrieve stakes separately as he comes to them. If need be, partial shovelful of concrete can be handed to the tamp operator to fill holes as needed.

Wet screeding

Wet screeding is done when forms are not available on one side of a slab to support a screed board. In addition, screed forms might not be available because of space limitations, the amount of work involved to place them, or any number of other reasons. Because they have plenty of experience, many professional concrete finishers prefer to wet screed along house and other structure walls when pouring 3 to 4-foot-wide walkways. The reason is simple— they can achieve an excellent screed without having to perform the extra work involved with screed-form construction.

To wet screed, first snap a chalk line against the wall you will pour next to. This will designate the concrete's top surface. In addition, snap a line 3½ inches above the first line to use as a reference for the top of your 2-×-4 screed board. Be sure that the lines are straight and conform with the grade you have established.

While pouring concrete, use a hand float to position concrete along the wall, extending out about 8 inches to a foot. Use a hand float to create a smooth flat strip of concrete with its edge on the lower chalk line (FIG. 9-16). It is this smooth and even strip of concrete that will serve as a screed board guide along with the upper chalk line. Work on this strip until you have it correctly placed for about 6 to 8 feet along the wall.

You will have to direct all of your attention on the chalk line and flat concrete strip at the same time while working with a wet screed. Because you cannot easily look for high and low spots, have the screed board worker on the other end of the board and the helper take care of them.

Use a firm hand to glide the screed board over concrete while following the upper chalk line. With two guides to work from, there should be no significant problems (FIG. 9-17). The sawing-type motion suggested for

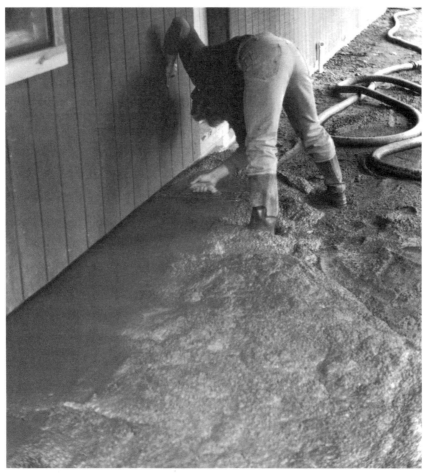

9-16 Concrete is floated up to the designated chalk line for a distance of 8 to 10 feet before actually screeding. A lip created by the plywood wall siding helps this finisher to accurately float concrete into place.

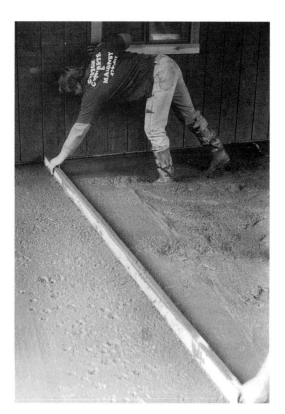

9-17 Operating the free end of a wet screed requires total concentration.

regular screeding cannot be used when wet screeding. That much board movement makes it almost impossible for a wet screeder to maintain a flat screed posture.

As with other screed work, you have to go back over areas that have divots, low spots, and other imperfections. Do the job right, no matter how many attempts it takes. Although wet screeding is easiest on narrow walkways and small slabs, professional finishers have had good results wet screeding up to 8-foot widths. Their success in this endeavor is based on experience, patience, and excellent screed worker communication. Novices are advised to rely on screed forms for any job wider then 3 feet.

OVERVIEW

Whether your job is screeded on top of form boards, screed forms, or just wet screeded, the end result is what's important. Pouring concrete and screeding it off requires a lot of physical labor. By the same token, these concrete placement operations will determine, to a great extent, how well your concrete pour turns out. No amount of finish work can remove excessive high or low spots. Once you call the screed job good, that's it. All the rest of the work will involve smoothing the surface and making it look nice. Therefore, have plenty of help available for the pour, fully explain the job at hand, do not be afraid to stop the flow of concrete to catch up screed operations, and do not get in an uncontrolled hurry. If the mud starts to dry out, add water. If a form breaks loose, stop and repair it. By all means, take care of problems immediately or direct someone else to do repairs for you. Have all of your equipment ready at hand, including screed boards that fit across forms adequately. If the job includes a narrow walkway and a larger patio slab, have a separate screed board available for both so you won't have to hassle with a long screed board just to screed a narrow walkway.

Chapter 10

The first hour

*T*he first hour after concrete has been poured and screeded will be busy. Along with tamping and floating, you might have to insert stirrups or J-bolts, anchoring devices used to support patio cover posts or walls. In addition, you'll have expansion joints, edges, and seams to work on; screed forms to pull; and overflow concrete to remove from adjacent, preexisting slabs or yard areas. Some cleanup operations, such as rinsing wheelbarrows, shovels, and screed boards, can also be started. Although thin layers of concrete residue will harden faster than slabs, you can delay tool and equipment cleaning until after tamping, bull floating, and other initial concrete finishing chores are complete. However, if plenty of helpers are available, have one start cleanup operations early on to be sure that concrete residue is completely washed away before it sets up.

TAMPING

Optimally, tamp work should start as soon as concrete has been screeded. With enough helpers on site, one could begin tamping right behind screeders. For all intents and purposes, tamp work can never be started too soon after screeding. In fact, the earlier you tamp, the easier it will be. This is because concrete is still wet enough to allow aggregate to be easily pushed into the mix. Waiting too long lets water and moisture evaporate from slabs to make the mix hard and, in turn, make it more difficult to force aggregate down from the surface.

Tamps are designed strictly for use on concrete surfaces to push down aggregate and bring up creamy concrete mud (FIG. 10-1). Wire mesh screens on the bottoms of most tamps usually measure 3 feet long by about 8 inches wide. Because of this relatively small size, tamp users have to walk around wet slabs in order to reach every square inch of the surface. Working in a backward direction, tampers fill in footprints by tapping around them with the bottom of their foot. Once initially filled in,

10-1 Drop tamps on fresh concrete to push down aggregate and bring up cream. This tamp has been operated correctly as indicated by the uniform pattern left behind.

they simply tamp over the spots to flatten and smooth them over (FIG. 10-2).

Fresh concrete is easy to tamp, as there is plenty of moisture in the mix to allow aggregate to be forced down and covered with a solid layer of cream. Conversely, concrete that has sat around for a while gets stiff and dry, and it will have to be forcibly pounded with a tamp over and over to achieve a smooth and rock-free surface—a very labor-intensive operation.

The same dilemma can happen if concrete is poured over a dry dirt base, especially on a hot day. All of the water in fresh concrete will be rapidly absorbed by dry dirt and quickly evaporated from the surface, leaving behind stiff concrete that is almost impossible to tamp with any success. To avoid that situation, be sure base material is thoroughly soaked with water and that tamping begins just as soon as screeding maneuvers allow.

Unless concrete is very dry, a great deal of downward tamping pressure is not necessary. The weight of a tool alone dropped from just a few inches above the surface is generally enough to provide a sufficient job. Done correctly, tamping results in a flat slab with no holes or protruding rocks but with a wafflelike surface finish. This pattern is created by the wire mesh screen on the bottom of the tamp. This wet, bumpy texture is erased and smoothed with a bull float (FIG. 10-3).

While operating a tamp, maintain the screen mesh in a flat plane,

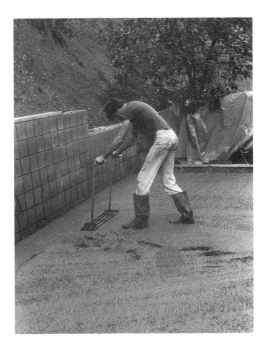

10-2 Footprints are unavoidable while tamping. Lightly tap footprints with your foot to fill in areas and then tamp over them to flatten surrounding concrete.

10-3 Wide bull floats do an excellent job of smoothing bumpy-textured concrete surfaces after tamping. Bull float as soon as possible to ensure concrete is wet enough to accept easy smoothing.

even with the surface. These tools are designed to make the job comfortable for the operator at any angle and easiest to apply in flat motions. If they are tamped against concrete surfaces with the front or back edge hitting first, a wavy and divot-marked surface texture will result. This type of inconsistent and uneven surface texture is difficult to flatten later with a bull float and fresno applications.

You can determine the flatness of your tamping efforts by evaluating the tamp's screen mesh pattern. If it is not evenly dispersed, adjust the position of your hands on the handles either toward the front or back until you start tamping evenly.

Screed work frequently leaves behind very slight ridges running across slabs in the direction boards were operated. To help flatten ridges and perfect surface flatness, tamp in the opposite direction of screeding. If you moved the screed board from east to west, then tamp from north to south. Tamping should start where the first concrete was poured, operated over 2 to 3-foot areas and then backward. Fill in footprints by tapping around them with your foot, then flatten spots with the tamp.

Once you have made a pass from one side of the slab to the other, simply turn around and work back the other way. You will tamp back and forth until the entire slab surface has been covered. Reach out as far as possible while keeping the bottom of the tamp flat to the concrete surface. Allowing one edge of a tamp to contact concrete harder or sooner than the other will cause minor divots on the surface, which are difficult to smooth later on. Tamps are designed with a balance factor to help users apply them flat and evenly. Let the weight of the tool do most of the work. Trying to force it into concrete makes the job harder to accomplish and usually results in uneven surfaces.

Many experienced concrete finishers have noticed best results along form edges when they apply tamps with their 3-foot front side in line with forms. This longer tamp edge is easier to guide next to forms and also does a better job of tamping. Do this along all form edges, including stringers.

If your concrete slab has suddenly gotten very hard and dry before tamping—because the base was not wet down sufficiently or it is extremely hot outside—you might really have to pound down on the tamp with every blow. This is done to knock rock and aggregate down, bring up all available moisture and cream, fill in tiny crack voids surrounding surface rock, and to create enough cream for finishing tools to work with. Should extra hard tamp pounding not work well, you might be forced into sprinkling a fine mist of water over the slab surface.

Although adding water to concrete finishes tends to dilute mixtures and might weaken surfaces, a very fine spray lightly applied to hardened surfaces is about the only way to make them soft enough to accept a decent tamp job. In these situations, you must have someone operate a bull float over the slab immediately after each tamp pass. If not, you will have an exceptionally tough time smoothing the wafflelike pattern left behind by the tamp. By all means, avoid this dilemma by soaking bases with plenty of water before pouring and by having enough helpers available so tamping can take place immediately after, or during, screed operations.

BULL FLOATING

Bull float tools are 3-foot pieces of wood or metal about 1-inch thick and generally measuring around 8 to 10 inches wide and are equipped with adapters for pole extensions because you cannot walk on slabs at this point. They are used to smooth wet concrete surfaces immediately after tamp work is completed and from various angles to fill in cat's eyes—small and shallow low spots that a bull float never touched. The weight and wide surface area of bull floats work together to smooth tamping bumps and bring up moisture.

Most professionals initially operate bull floats from a position perpendicular to that of tamping. In other words, if a screed board was pulled over a slab from east to west, tamping would run in a north/south direction to help flatten out any minor divots or dents produced by jagged screeding. A bull float is then operated in an east/west manner to compensate for any tamping flaws, filling in shallow low spots in the process. If need be, of course, you can maneuver a bull float in any direction to pick up cream and fill in low spots.

Bull float work is done immediately after tamping. In fact, if enough helpers are available, one should bull float while another tamps and others complete different tasks. The sooner tamp and bull float work is accomplished, the better. Both of these operations bring up lots of cream to the surface, and it is this cream that will ultimately be finished to give your concrete slab surface its final appearance.

Waiting too long to float makes smoothing jobs difficult. Instead of a lot of wet cream to work with, moisture will have evaporated leaving behind tiny cracks and pockmarks that cream has sunk into (FIG. 10-4). This condition requires you to bull float over the surface many times and from all directions, essentially working up cream from wet areas and floating it over to dry ones. All holes will have to be filled. The entire surface must have a somewhat smooth texture before fresno work can be done.

Because concrete should still be wet, lines and ridges will be left behind from bull float edges. If these lines and ridges are so prominent that they protrude more than a half inch, you should bull float a second time, about 5 to 10 minutes after completing the first float. Don't worry if the surface finish does not remain perfectly smooth. It is normal for some smooth, slightly bumpy looking patterns to appear as moisture evaporates. Fresno work will take care of that.

Dry concrete conditions require that fresno work begin immediately, just as soon as you are done with the bull float. There is no time to waste between operations. Likewise, you will probably have to hand trowel finish right after the fresno. In other words, fast-drying concrete requires immediate and constant work. As soon as one procedure is accomplished,

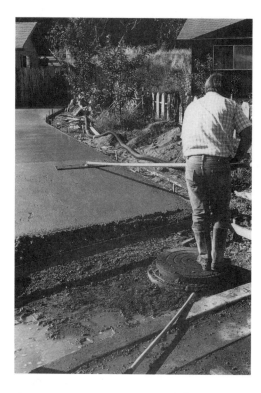

10-4 Concrete in this area has dried somewhat, making bull float work more labor-intensive. Operate bull floats back and forth across dry surfaces to work up cream, fill in holes, and smooth tamping marks.

the next one must start until all procedures have been completed and the slab is completely finished. Under normal conditions, though, you should have time to rest between bull float applications and fresno work and again after the fresno and before hand trowel maneuvers.

Because extension poles might have to reach up to 18 feet, sometimes more, you might have to make room for them around slabs or figure out alternative methods for applying the bull float and fresno. This is especially significant for slabs that are bordered on two or three sides by walls, fences, structures, etc. (FIG. 10-5). You might have to float and fresno half of a slab from one side and the other half from its opposite side.

Even if you have room to operate with more than three, 6-foot extension poles, you will have trouble. Poles extending over 18 feet tend to bow in the middle, making it almost impossible to raise them high enough so the back edge of your bull float or fresno can be raised off of the slab surface when the tools are pulled back toward you. Be sure to consider factors like this before the pour. Make arrangements to remove fence sections or make other adjustments so that all parts of your slab can be easily accessed by the bull float and fresno.

Adjust the bull float adapter swivel so that the tool is pushed away from you with its front edge off of the surface to prevent it from digging into the concrete. This will entail holding poles low to the ground (FIG. 10-6). The swivel adjustment should also be positioned so that you are

10-5 Slab areas bordered by obstructions might require bull float work from a number of different directions.

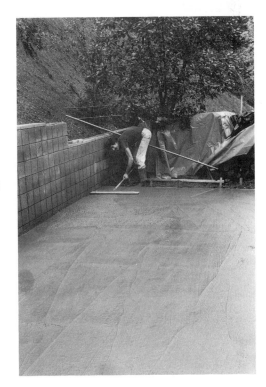

10-6 Heavy lines left on the surface as a result of bull float maneuvers are a sign that concrete is still wet.

able to raise poles high enough to lift the back edge of the float while pulling it back toward you so it doesn't dig into the concrete. Simply hold poles low while extending the float out onto the slab and then raise them high to bring the float back.

Ideally, bull floats should be operated over slabs in the opposite direction of tamping, just as tamping was done opposite of screeding. Working in opposing directions helps to better smooth and flatten concrete surfaces, as well as fill small low spots and other imperfections. Obstructions can prevent you from operating a bull float in the opposite direction of tamp work. This is common and you can certainly end up with a fine looking slab even though the tamp and bull float were operated in the same direction. It is just that working in opposite directions can provide a little better coverage, like mowing lawns in one direction and then the other to guarantee every blade of grass is cut.

Staying ahead of the concrete is the name of the game. At no time will you want concrete to dry out and get hard before you are ready for final hand trowel finishing. Pay close attention to timing and start each successive work phase right away. This first hour of working concrete will greatly affect how smooth the rest of the work goes.

HAND FLOATING

Hand floats are small trowels made of wood. Like bull floats, they are used to initially place and smooth rough concrete. Steel hand trowels are used to actually effect final finishes. Hand floats are useful in placing mud into tight spaces while pouring and screeding, flattening out rough spots, and smoothing surface material. The front portion of the tool works great as a scoop to place small amounts of concrete under mud sills, lips formed by wall siding, and lots of other limited access spaces. While screeding, many finishers keep a hand float nearby to assist in moving excess concrete away from screed boards.

Form and stringer tops are quickly cleaned by scraping a hand float edge against them (FIG. 10-7). You can also use this tool to remove concrete residue from posts and other obstacles inside a concrete area. Hand floats work much better than your hand to smooth and reposition concrete simply because they are flat.

Hand floats are often used to smooth rough, wet concrete along slab edges close to forms (FIG. 10-8). Bull floats are big and not always easy to maneuver along edges. Slab perimeters generally dry quicker than the middle because moisture is lost at the sides as well as the tops and bottoms. Therefore, hand floats might need to be used along slab edges to scrub concrete and bring up creamy finishes. Hand float wooden blade textures work great to force down aggregate and bring up workable cream slurries. Once a dry area has been scrubbed to a creamy finish, a couple of light hand float swipes should smooth it in preparation for a fresno (FIG. 10-9).

Places too difficult to reach with a bull float might be accessed with a hand float. You will not be able to walk on concrete after it has been

10-7 While other work is being tended to, extra helpers can clean form tops with a hand float for a clean working guide while edging.

10-8 Scrub and smooth rough concrete along the edge of forms with a hand float.

10-9 Once rough concrete has been sufficiently scrubbed to bring up cream and fill holes, a few light passes should smooth it with an even texture.

tamped, but you might be able to reach out-of-the-way spaces by walking along walls, forms, steps, or fence rails. Hand floats are miniature bull floats. Their design and construction is intended to offer the same results as bull floats but with much greater maneuverability (FIG. 10-10).

The perimeters around large obstructions in the middle of slabs are easily smoothed with a hand float. Items such as manhole covers, water meter lids, and so on, are difficult to work around with large bull floats. Hand floats are used for a lot of various tasks. While checking and smoothing slab edges, clean off excess concrete from the tops of forms so that mud does not accidentally get knocked onto slabs during other work.

SEAMING

Seams, also called control joints, are made on concrete surfaces in hopes that any eventual crack will occur inside the seam rather than diagonally across nice finishes. Although seams are most common on walkways and sidewalks, large driveways and other big slabs should have them too. Used in conjunction with expansion joints, seams provide a valuable means by which to control cracks.

Walkways should have seams about every 3 to 4 feet with an expansion joint every 10 to 12 feet but it depends on the walkway length. Seams should be placed so that they are evenly spaced and symmetrically

10-10 Because of the round porch, a regular bull float was not able to fill in concrete around bricks or smooth in tight spaces. A hand float is used to accomplish what the large bull float could not.

patterned. Seams and expansion joints are needed for walkways because of their generally long and narrow shape. Their length does not allow room for concrete expansion on hot days and the narrow width doesn't provide enough concrete to add necessary strength or support.

Driveways should be equipped with a seam every 10 feet and an expansion joint every 20 feet running across their width. Unlike patio slabs that usually do not have to support a great deal of weight, driveways are consistently subjected to heavy vehicles. Heavy weight loads increase the chance of cracks and seams help to control this.

Seams are placed using a 2-×-4 board as a guide. The board should be long enough to span the slab, preferably placed so that it is supported on both ends by forms. Nails strategically placed along outer form faces are great for designating the location of seams. This should be done before the pour when you have plenty of time to measure the job and determine the most accurate spacing.

Seamers have a definitive ridge in the center of the blade. This ridge is what causes a seam groove on concrete surfaces. Because this ridge is located a few inches away from the tool's outer edge, you must position 2-×-4 guides appropriately. In other words, if you put a seam at the nail guides, you must position your 2-×-4 guide away from the nail far enough to account for the distance between the seamer ridge and the outer edge. If the 2-×-4 edge is placed in line with the nail, the seamer puts the seam a few inches away from the nail's position (FIG. 10-11).

10-11 A long, 2-×-4 board is used as a guide for an initial seam installation. Because this driveway is so wide, a walking seamer is needed.

Seams are easiest to apply while concrete is wet. The soft texture of wet mud allows seamers to smoothly glide across surfaces to quickly create deep, crisp seams. Waiting too long to effect seams makes pushing aggregate down and forming cream into a crisp groove difficult. Therefore, plan to initially install seams right after concluding bull float maneuvers.

After making seams, remove the 2-×-4 guide boards. You might need help on long ones; they sometimes stick to wet slabs with a suctionlike grip. Use a bull float to remove lines or imperfections left by the board. After an initial seam has been made, don't worry about it filling with cream. Your main purpose was to move rock and grit out of the groove. Fresno activity will fill seams with cream, which is easily reshaped with subsequent seamer applications (FIG. 10-12). Plan to run seamers along seams after every other fresno application.

You should not have to use a 2-×-4 guide every time you use a seamer. Guides are only necessary while initially making grooves (FIG. 10-13). After the first application, an indentation has already been established and cream will not cause seamers to wander like rocks would. Take your time and maintain a firm steady motion. Lines to the sides of seams left behind by seamer edges are wiped out with a fresno or hand trowel. As the slab hardens, use a seamer to simply finish groove surfaces as opposed to actually shaping them. A light touch is best. Once you're done, be sure to wipe out edge lines with a trowel.

Walkway seams are easily made using a hand-held seamer. For longer

10-12 Fresno applications done after initial seams have been made will fill in grooves with cream. A second seam application will not be hindered by aggregate and cream will conform to the desired shape.

10-13 Because aggregate was pushed down early, this second seam application can be carefully completed without a 2-×-4 guide.

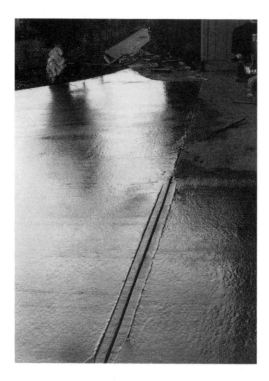

10-14 A corner has been formed by pouring around a preexisting concrete apron in front of a garage. This seam should control cracks in this area. It is important to install seams and/or expansion joints anywhere corners are formed.

seams on driveways or sports courts, you'll need a walking seamer. These larger tools are available at rental yards. Their adapters should conform to those on your bull float and fresno so that the same extension poles can be used on all three tools.

In addition to placing seams at strategic points across slabs and walkways, you should insert them at every corner (FIG. 10-14). Each time a walkway makes a turn, the corner of it should have seams inserted out from its inside corner. If seams or expansion joints are not provided, that corner of the walkway will crack (FIG. 10-15).

EDGING

Edging wet concrete soon after bull floating has the same effect as putting seams in early; it gets rocks out of the way while mud is still wet and soft. And, like seams, these edges will probably get filled with cream from fresno operations. Plan to go over edges after every other fresno application.

Walking edgers make quick work of edging concrete (FIG. 10-16). The tool's rounded side is inserted between a form edge and concrete. Its flat part will smooth concrete away from the slab edge while the rounded section actually effects a curved corner along concrete edges. As with other concrete tools, you have to keep the front part up and out of concrete while moving forward and the back part up when going backward (FIG. 10-17).

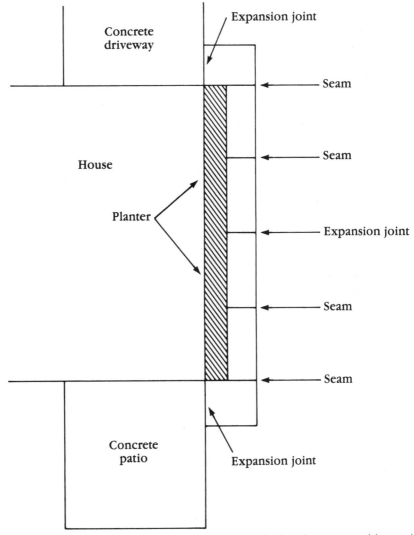

10-15 A walkway poured against a patio slab in the back and a concrete driveway in front should be equipped with seams and expansion joints as illustrated.

Forms serve as guides while operating edgers. Gently allow the side of your edger to run against forms while keeping a close eye that the tool does not stray off course. Concrete's newly formed round edges should be smooth, but go over them again if holes or blotches appear. Even though this might be just an initial edging, try to shape and smooth edges correctly. This way, later maneuvers will be easier to effect and finish.

Hand edgers must be available for most concrete pours even though a walking edger is on site. These are used in tight spots where walking models might not fit. They are also great for repairing edges that might

10-16 While moving forward, be sure an edger's front edge is raised off the surface.

10-17 Operated smoothly and with balanced pressure, hand edgers do a fine job of making crisp rounded corners. Remember to keep the leading edge raised off the surface to avoid making divots on the surface.

eventually suffer nicks, dents, or other blemishes. Hand edgers are much more maneuverable than walking ones and better suited for intricate work.

If you do plan to use both types of edgers, make sure they both have identical, rounded-arc dimensions. Edgers are generally available in 1/4-inch, 3/8-inch, and 1/2-inch arcs, with the 3/8-inch model seemingly the most popular.

Hand edgers are essentially operated the same as their larger counterparts. You must raise the front end while moving forward and the back end when going backward. Apply even pressure to edgers. Too much force on a rounded side tends to push an arc's edge into a slab's side, making an awkward concrete edge. Too much pressure on the edger's flat sides cause deep lines on concrete just off the tool's outside edge. Evenly applied pressure effects crisp, round concrete edges with hardly any line from the flat side.

Lines left on concrete surfaces are wiped out with hand trowels. However, extra deep lines continually made with each edger pass require significant amounts of trowel work. This is not good, because grit will eventually be worked up onto the surface as edges become more uneven (FIG. 10-18).

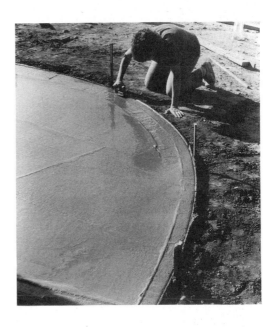

10-18 The purpose of an early initial edging is just to make sure aggregate is pushed down and out of the way. Overworking edges results in heavy lines that must be hand floated while concrete is still wet.

All concrete surfaces that come in contact with forms, expansion joints, and preexisting concrete flat work should be edged. This includes both sides of stringer tops as well. One of the few times when edges are not needed are along house and other structural walls. Concrete is finished flat against them.

FIRST FRESNO

Your first fresno application should be made when a concrete surface has lost most of its watery sheen (FIG. 10-19). Metal-bladed fresno tools are basically 3-foot-long finishing trowels that bring moisture up to slab surfaces while further smoothing an existing finish with each pass. For the most part, first fresno applications are done about 10 to 20 minutes after bull float work is complete.

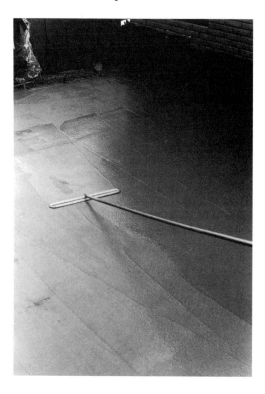

10-19 Darker, shinier concrete is a sign that moisture is evaporating from the surface but that fresno work is bringing moisture up. Lines left by the round fresno blade indicates that this concrete is still wet.

Although fresno passes made too early might bring up excessive moisture, leave ridges behind from tool edges, and possibly bring up some sand particles, it is better to fresno too early than too late. If you are in doubt as to how long to wait, try the fresno on a slab section to test the results.

A good fresno timing indicator is the watery sheen on slab surfaces. Once the sheen begins to dissipate, it is usually time to fresno. On very hot days, you should apply a fresno right after bull floating. Hot weather causes concrete to flash, which means it dries out quickly and suddenly. Cool, damp weather tends to keep slabs wetter for much longer periods (FIG. 10-20). Sometimes, cooler conditions allow finishers to wait up to an hour before using a fresno.

When applied at just the right time, fresno applications smooth slab surfaces considerably. As moisture levels in concrete start to drop, how-

10-20 Fresno applications are made as watery slab surface sheen's disappear. As concrete gets harder, fresnos do a better job of smoothing concrete.

ever, surface textures once again take on a rougher appearance. Each succeeding fresno application further smooths the surface. In fact, it is not uncommon to fresno slabs four and five times.

Operating a fresno with extension poles can be awkward, so take care not to trip over obstacles or knock things over with the poles. Most finishers simply stand next to forms and use a hand-over-hand method to push a fresno across a slab. To pull it back, they raise poles and walk backward. You could pull it back with just your arms but that jerky motion will result in surface imperfections. Walking backward works best.

A fresno is operated just like a bull float. You must hold poles low to the ground while pushing it away and up high when pulling it back toward you. The front of the tool must be raised while moving forward and the back edge must be raised while being pulled back.

Fresno adapters are also adjustable. Make any necessary adjustments to the swivel before you start. On long slabs, be absolutely sure the blade is adjusted so that you really have to hold poles low to get the fresno out and away. This way, you should have no problem raising it high enough to bring it back without its back edge scraping on the concrete. You can get yourself in a heck of a bind if the blade is adjusted too high and you can't raise the pole extensions high enough to get the fresno off the slab without using a stepladder.

Fresno adapter swivels generally allow for sideways adjustments as well as for up and down directions. This allows you to use a fresno at an angle in tight spots. An example would be a slab poured next to a fence. If the fence blocks the extension poles coming straight off the slab, maybe there would be room if they came off at an angle. Using a fresno at an angle works great while walking along and finishing a 3-foot, 1-inch walkway lengthwise.

Chapter 11 describes successive fresno applications thoroughly. For now, though, remember to apply your fresno in a smooth and steady motion with the front edge up while moving forward and the back edge raised while pulling back. As long as your slab does not flash and the weather is a mild 70 degrees, you should be able to stop and catch your breath after the first fresno—providing all of the other, previously described chores have been completed.

MISCELLANEOUS TASKS

Removing screed pins or stakes can be difficult. Although designed to be pulled out easily, you can pound them into the ground so hard they resist easy removal. The best way to loosen them, and also the best way to prevent wrenching your back, is to smack them with a hammer. Hit them on one side and then the other to wriggle them loose. On soft soil bases, some finishers easily loosen pins and stakes with a shovel (FIG. 10-21).

Cleaning forms and tools is a task that should not go unattended. Water applied at a high pressure usually works great. Just be sure water does not get sprayed on new concrete. Use a stiff brush or foxtail broom to help loosen and remove stubborn patches of concrete residue.

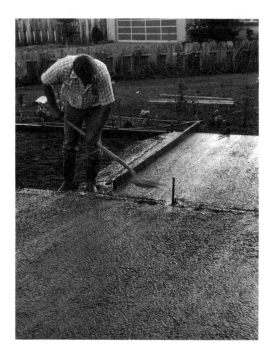

10-21 In soft ground, you might be able to reach out with a shovel to loosen screed pins or stakes. Most of the time, though, you'll need to use a hammer.

Screed boards, screed forms, screed pins, stakes, and tools should be cleaned as soon as possible. Wet concrete is easier to remove than dried on remnants. Tools that will be used again must also be cleaned between applications. Hand trowels, seamers, and floats can be placed in a five-gallon bucket partially filled with water. Use a brush to clean them. Be very careful of the sharp edges on steel trowels and edgers. They can become razor sharp in no time after being "honed" on concrete.

During the first hour, stirrups and J-bolts should be installed. You have already learned how to designate their prospective locations (nails on the sides of forms). When placing more than one, use a string as a guide to make sure they are inserted in line with each other. Install them right after the bull float while concrete is still soft and wet. You might have to use a hammer to help persuade them into position. Recheck their position after installation to be sure they are straight up and down and in line with each other.

During the first hour, 2-inch-wide benderboard expansion joint strips must also be installed, providing you planned for them instead of felt or stringers. Place them after the first bull float. Use a hand float to smooth concrete around them afterward.

By the time you get all of your projects completed, it will probably be time to apply a second fresno. The first hour of any concrete job is always a busy one. Maybe you should have a list posted at your job site describing all of the tasks you plan to accomplish? This check list should ensure you don't forget anything and might serve you well as an overall guide.

Chapter **11**

Basic
flat work
finishing

*B*asic flat work finishing entails work with a fresno, edger, seamer, and metal finishing trowels. At this point, providing previously described preliminary work has been completed correctly, your primary focus will be strictly confined to slab surface smoothing. Concrete should have already been properly placed, screeded, and tamped. In addition, first edges, seams, and a first fresno should have been applied. Also, extra installations such as expansion joints, stirrups, and J-bolts should have been taken care of.

Remember that all outdoor concrete flat work must provide some sort of traction when it gets wet from lawn sprinklers, rain, washing, and so on. The easiest and most common method of providing concrete surface traction is by applying a broom finish. Along with providing adequate traction for wet concrete, brooming tends to cover and hide many small finishing imperfections.

Be prepared to apply this finish by having a soft-bristled push broom handy. If your slab is so large that you will not be able to reach all parts of it with your broom's regular handle, have it set up to accept bull float and fresno extension poles.

FRESNO FINISHING

Since you should have applied a first fresno during the first hour after pouring concrete, you should have a feel for how the tool is operated. Successive applications are no different except that you might have to use increased downward pressure to smooth concrete that has gotten harder. Most of the time, additional pressure is not needed until the third or fourth fresno. However, very hot weather can cause surfaces to flash, requiring extra power for smoothing.

Generally, second fresno applications are made when the watery sheen has once again left a slab's surface. The very top layer of concrete could still be wet and creamy, but its visible surface will have experienced some moisture evaporation and appear dull. Ripples on the surface of slabs, which are smoothed out with fresno applications, are an indication of moisture evaporation.

Run your fresno across the entire slab. As you do this, you'll notice that fresno-covered finishes become darker than untouched parts. This is normal, and is caused by the fresno's steel blade bringing up moisture from just under the surface (FIG. 11-1). As long as your fresno glides easily over slabs and brings up moisture, you are assured that the surface is still wet. Although it is not always good to overwork slabs, you should attempt to apply a fresno every time surfaces loose their sheen. If you start to fresno a dull surface only to find that the slab is still ''soupy'' (very wet and creamy), just stop with that pass and allow concrete to set up for another 10 or 15 minutes.

If, however, the slab loses its sheen and you neglect to fresno, especially on hot days, the delay could cause fresno applications to create pockmarks, slight holes made when bits of dry concrete stick to the blade or by excessive cream moisture evaporation (FIG. 11-2). To fill in these holes, you must operate a fresno across the slab a number of times. Bring cream and moisture to those spots by maneuvering the fresno at angles to holes so cream can be brought in from surrounding areas. Work the

11-1 The light tinted concrete in front of the blade shows that this spot has yet to be touched by the fresno blade. Notice the darker surfaces that have already been smoothed by the fresno—indicating that cream was brought up.

fresno back and forth over bad spots using whatever downward pressure is necessary. This added pressure on the fresno helps to scrub slabs and bring up moisture. Once an area is scrubbed, run the tool over it lightly to effect a smooth, even finish (FIG. 11-3).

As with previous bull float and fresno applications, you might have had to remove an extension pole or two in order to reach tight spots. Work the tool back and forth to create a smooth finish. This is really

11-2 Run the fresno back and forth over rough surface spots to bring up moisture and cream to fill small holes and pockets. If necessary, operate tools from a different direction.

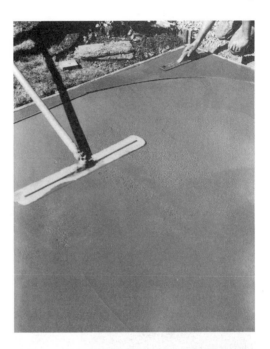

11-3 A fresno wipes out scrubbing maneuvers with one smooth gliding pass. Make the last pass a full one; side to side in one nonstop motion.

important for low spots (cat's eyes). These areas will obviously appear as spots that the fresno did not touch. In fact, they were not touched, as evidenced by the darker concrete color surrounding them. Work the fresno around these areas like you would for holes or pockmarks. Sometimes, slight twisting pressure applied to poles will cause one side of a fresno to slide into shallow low spots to finish them. Use this maneuver with scrubbing action to bring up cream and fill low spots (FIG. 11-4).

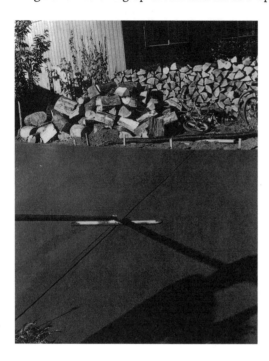

11-4 A low spot is located just behind the left side of the fresno blade. As the tool passes over the spot again, the operator will twist the handle to force extra downward pressure on the left side.

The more concrete sets up and hardens, the less a fresno will affect smoothing and finishing. When concrete is just about ready for hand trowel finishing, a fresno will make a distinctive ringing sound—in effect, metal rubbing against a hard surface. Fresno as best you can with downward pressure. A lot of finishers add weights to the tops of fresno blades to help them finish harder slab surfaces. These weights actually work in unison with downward pressure applied to extension poles. The added force of weights allow finishers to fresno slabs one last time before having to get out on slabs to hand finish.

Not all slabs require the use of fresno weights. More of a luxury than a necessity, weights simply let finishers get slabs a bit smoother before having to hand finish. The smoother a slab is before hand trowel operations, the easier it is to hand finish (FIG. 11-5).

Most concrete slabs require at least three fresno applications. Timing is important. Although you should wait until watery sheens have gone, do not wait longer than 20 minutes in summer and 30 minutes in winter to fresno again. It is much better to apply an extra fresno than to wait too

11-5 Weights on this fresno are working to bring up moisture and smooth the surface over an installed seam. This will be the last fresno application, as concrete is just about hard enough to support knee boards and hand trowel finishing.

long between applications. Quality fresno work will go a long way toward reducing the amount of hand trowel work required for professional finishing results. The work accomplished with a fresno is work that will not have to be done by hand.

Deciding when a fresno is no longer useful can be tricky. The best you can do is watch and listen. Look to see how much actual smoothing a fresno is doing and listen for the distinctive metallic ringing sound made by blades rubbing against hard surfaces. This ringing noise signals that a slab is about ready for hand trowel finishing. To guarantee that concrete is ready for hand trowel finishing, test areas along edges next to forms. If you are able to effect a really smooth and blemish-free finish with metal hand trowels that also make a metallic ringing sound, the slab is ready (FIG. 11-6).

A final word about fresno work. Too many novices are afraid of ruining slabs with fresno applications. It is true that excessive fresno work will bring up sand and grit and too few fresno passes will make hand trowel work really labor-intensive. But, if you use common sense and realize that the tool is simply designed to reduce hand trowel work, you should be able to use it to its full advantage.

Obviously, if you push and pull a fresno across a slab and do nothing but move soupy cream from one side of the job to the other, you are wasting time and valuable cream. Take a break, clean some tools and let concrete set up. On the other hand, if a fresno does a good job of smoothing

11-6 Begin hand trowel finishing along slab perimeters, which can easily be reached from outside forms.

the surface, brings up moisture as determined by surface color, fills in small holes and smooths imperfections, it is working as intended.

As long as the tool smooths concrete and improves texture, great, keep using it. If only one spot is giving you trouble, work on it. Once a fresno fails to make much of an impression, it is probably time to grab your knee boards and hand trowels.

HAND TROWEL FINISHING

Most concrete slabs start setting up along their outer edges first. Therefore, test finish readiness along perimeters first (FIG. 11-7). Use one trowel to lean on and the other to reach out and work the surface. By reaching out as far as possible, you eliminate having to use knee boards in those areas. Proceed around the entire slab, leaving just a space for entering and exiting the middle (FIG. 11-8). Although most hand finishing starts where concrete was first poured, and likely the place quickest to harden, other finishers might spread out along edges to finish other sections.

Operating hand trowels does not require a great deal of experience. Although professionals can do this work much faster and with fewer mistakes, novices can accomplish the same goal by taking their time and concentrating on the overall objective—a smooth, line-free, hole-free surface. Pool trowels (those with rounded front and back sides) are highly recommended for novice finishers. These tools do a great job of finishing and

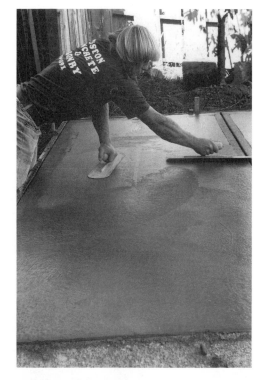

11-7 The first concrete poured is usually the first to harden; it has had more time for moisture to evaporate.

11-8 Long, wide smooth hand-trowel passes work best to smooth concrete.

greatly reduce the formation of prominent lines left behind by regular rectangular hand trowels.

Hand trowels are swiped over concrete in half-circle motions. Stretched out, this is a natural motion for your arm. Look at finished concrete slabs, like your garage, and you will notice definite trowel passes that were made in an arc shape. As you had to do with all of the rest of the finishing tools, you will have to raise the left-hand edge of your trowel when moving toward the left and the right side when going right. This prevents the leading edge from digging into the concrete surface. You will only make the mistake of not lifting a leading edge once, as the damage done by this will be more than obvious.

A rounded trowel motion is not mandatory for all finishing endeavors, however. If a slab you are finishing is quite small, like a stoop or step, you can simply apply a trowel on one side and maneuver straight across to the other (FIG. 11-9). This is especially true for slabs with unique designs or small width sections. When forming mowing strips or other extra narrow jobs, try to make them wide enough for a trowel to maneuver inside them (FIG. 11-10).

11-9 Finishing trowels are not always maneuvered in an arc motion. On small slabs, they can be operated straight from one side to the other. Here, a trowel edge is lightly placed next to the doorsill and then moved across the concrete toward a form.

As important as it is to raise the leading trowel edge, it is just as important to maintain trowels at workable angles. Raised too high, a trailing edge will scrape against concrete, causing tiny rock tips to pop up and cream to be scraped off. If the leading edge is too low, cream will swirl

11-10 When forming borders, mowing strips or any flat work design with tight dimensions, leave enough space to maneuver tools.

under trowels, leaving behind unacceptable rough finishes. Practice with hand trowels to get a feel for them, their correct application angle, and a hand position most comfortable and functional for the job.

Moving hand trowels too slow will not effect smooth finishes. Going too fast usually results in mistakes and blemishes. A steady controlled speed normally works out best. Move your arm about as fast as you would while operating a paint roller on a wall. You could go faster or slower but only practice can determine that.

Attempting to finish concrete that is still too wet will result in trowel lines and soupy ripples (FIG. 11-11). In these situations, use a fresno to see if its wide blade will do a better finishing job. Hand trowel work at this point is a waste of time. A wet slab will not hold a smooth finish. Conversely, if your trowel doesn't seem to do much to smooth a hardened surface section, you'll have to scrub up some cream and then apply a finishing swipe or two (FIG. 11-12). Use your smallest metal hand trowel for scrubbing. Vigorously rub it back and forth across the rough spot until a layer of cream is brought up (FIG. 11-13). The entire scrubbed area must have the same texture and swirl-type finish as the rest of the slab. Once this is accomplished, all holes and pockmarks should be filled.

Only very rough spots need to be scrubbed. Most other areas cream-up with just two or three vigorous trowel passes. Work small areas at a time with your scrubber trowel to bring up moisture and cream. Then,

11-11 Trying to finish concrete that is still too wet will result in lots of trowel lines and a soupy surface texture.

11-12 This is a rough concrete surface that needs to be scrubbed and then finished.

make a swipe or two with your larger trowel to finish the area off. Don't worry about very light, fine lines left behind by trowel edges. These will be wiped out by broom bristles later on.

Hand trowel finish work consists of two separate actions: scrubbing to bring up cream and moisture and finishing to smooth it all out (FIG. 11-14). One of the biggest mistakes novice hand trowel finishers make is

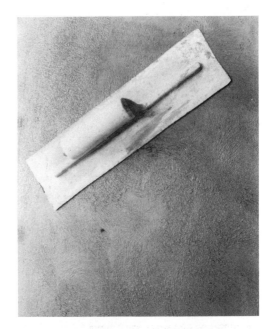

11-13 A rough but uniform texture shows that moisture and cream have been brought up to fill small holes.

11-14 This finisher reaches out to finish as much of the slab as can be reached from outside forms.

working concrete too much. Instead of just briskly working up some cream and then wiping it off, they continually scrub and finish. Scrub relatively smooth areas with three or four passes and then try to effect a final finish. Concerned that this effect will not be good enough, run a broom over it to really determine final effects. If it looks good, leave it and continue on.

Concrete work is no different than anything else when it comes to the old adage that practice makes perfect. If you are concerned about your concrete finishing ability, find a somewhat out-of-the-way section of concrete to practice on while it is still wet. Find a remote corner or section that will be covered with patio furniture or otherwise heavily broomed. Remember that finishing wet concrete results in a lot of lines left behind by trowel edges and, since wet mud is not hard enough, the finish will not be as expected. No matter. This practice helps you get a feel for the tool and to determine where your hand position on the tool works best.

Plan to finish center slab sections after successfully finishing around the perimeter outside of forms. Stretch out while leaning on a trowel, you might be able to reach much more of the surface than expected. This can really eliminate a lot of unnecessary knee board activity in the middle of slabs, while reducing your workload in the process. All areas inaccessible from the outside will have to be finished on knee boards.

11-15 This finisher is working backward on knee boards, finishing the concrete in front of him.

Use knee boards in a leap frog fashion working backward as you go. This way, you can trowel out impressions made by knee boards while progressing to another section (FIG. 11-15). The ultimate goal in finishing with knee boards is to finish the greatest amount of area while moving knee boards the fewest number of times. Reach out as far as practical from every side, because each time a knee board is placed on fresh concrete, some type of impression is made. Those impressions require scrubbing before finishing (FIG. 11-16). The more you move knee boards around, the more impressions you'll have to scrub out.

Along walls, especially rough textured ones, you might have to operate a trowel somewhat like a putty knife. Small sections of concrete might be very rough because a fresno could not adequately reach them. Use the tip of your trowel to initially pat concrete and then gently but firmly slide it away (FIG. 11-17). If more pressure is needed to wipe out extra rough spots, place the fingers of your free hand on top of the trowel tip and push down. Sometimes, a little concrete can be scraped off of walls and used to fill in holes and rough spots along them. Fill in as needed and then start the trowel swipe against the wall and move out. When you get toward the end of your reach, start lifting up on the trowel until it completely comes off the surface in one smooth motion; sort of like an airplane lifting off the ground.

Making hand trowel finishes absolutely perfect is only critical when finishing slabs that will not be broomed or otherwise custom finished, such as garage floors and tennis courts. Tennis courts are generally large pours that require a lot of forming and finishing expertise to perfect exact

11-16 Knee boards make an impression on concrete surfaces. Scrub knee board blemishes and then wipe them out with a finishing trowel.

11-17 Finishing concrete against walls might require extra trowel pressure. Push down on the end of a trowel with your free hand to help smooth rough spots.

grade standards and surface textures. Ordinary outdoor flat work requires a broom finish, at the least.

FINAL EDGING

You can go around slab perimeters with an edging tool after each fresno application but generally, you only need to after every other fresno (FIG. 11-18). Naturally, as concrete hardens you must maintain edges in excellent condition at all times, especially if fresno or trowel work results in lines or accumulations of cream.

Edgers must be kept flat on the surface in a side-to-side plane, raising the front edge to move forward and the rear edge to go backward. Too much pressure applied to the rounded side of the tool will cause edges to tilt down from form tops (FIG. 11-19). This results in the extreme outer perimeters of slabs to slope down slightly from the surface. On the other hand, too much pressure exerted on an edger's flat side edge causes deep lines (FIG. 11-20). For the most part, these lines are not major problems but simply create more work to finish a slab surface (FIG. 11-21). Therefore, it is a good idea to practice positioning and operating edgers so that the extreme outer edge does not slope or create deep lines. Although lines are wiped out with metal finishing trowels, the deeper and more frequent they appear, the greater chances are that they will turn into permanent surface creases.

11-18   Final edge finishing is done using edgers.

11-19   Excessive force on an edger's rounded side causes slab edges to slump down below form tops.

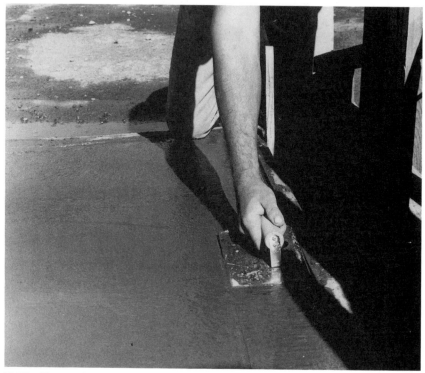

11-20 Too much pressure on an edger's flat side results in heavy lines that have to be wiped out with a finishing trowel.

Rough edges resulting from fresno work and moisture evaporation can be made crisp again with continued edger work. Try to minimize the amount of times you have to work edges to make them perfect, but do make as many passes as necessary to accomplish the job. This might sound contradictory, but the more you rub slab surfaces with finishing tools, the greater chances are of working up sand or grit. For rough edges, short back and forth strokes, steady and firm, will bring up cream and fill holes. When the area has been filled with cream, apply one or two light strokes to finish it off crisply and uniformly.

Most of the time, concrete finishers carry their edger tool with them while using hand trowels to finish perimeter slab areas. This way, they can finish edges and perimeter areas at the same time. Because trowel work can mark edges with lines or extra cream, do it before edging. After edges have been finished, apply a few light trowel strokes to wipe out inadvertent edger lines. By working on both perimeter finishing tasks and edges at the same time, finishers accomplish jobs quicker with less moving around. Conversely, should one helper be edging and another finishing, they might just get in each others' way.

Edging along stringers is just as important as edging around perimeter forms. For stringers in the middle of slabs, use a walking edger equipped

11-21 A rounded pool trowel works great for wiping out edging lines. Run a trowel lengthwise to knock down lines and then operate it from the form edge out to the finish area.

with extension poles. To effect final edges along stringers, bring your hand edger along while leapfrogging around the slab with knee boards and hand finishing trowels. As you reach stringer sections, edge along them as you would outer forms (FIG. 11-22).

Edging work must also be completed along expansion joints located in the middle of slabs, whether they are felt, stringer, or benderboard. Use a walking edger, as needed. Be sure to fill in holes along edges and finish areas next to joints with a hand trowel (FIG. 11-23). Apply the edger tool next to expansion joint edges to make them crisp and perfect (FIG. 11-24). Wipe out any edger lines and then finish surface areas with your trowel as far out as you can reach. By doing both finishing and edging chores at the same time, you accomplish all of the necessary work in that area while only having to place and work off of one knee board at a time.

Notice that in FIG. 11-24, the finisher edged both sides of the expansion joint at one time while positioned on just one side of it. This eliminates the need to edge opposite expansion joint sides when finishing concrete on that other side. By edging both sides of expansion joints at the same time, you are guaranteed that concrete will be the same consistency and hardness for both edges.

Final seam finishing is done much the same way as edges. After a final fresno application, run your seamer tool along grooves to effect crisp seams, including the base of grooves as well as the sides. Don't worry about lines made by the tool's outer edge; they'll be wiped out with a

11-22 Each section of the slab should be completely finished while you are there. Bring an edger or seamer to finish around obstacles, as is being done along this stringer.

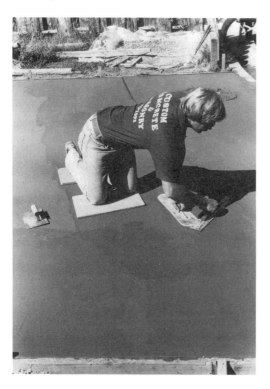

11-23 A benderboard expansion joint has nearly been covered with cream from repeated fresno applications. This finisher will use a hand-held edger to complete work on the joint after trowel finishing surrounding concrete within reach.

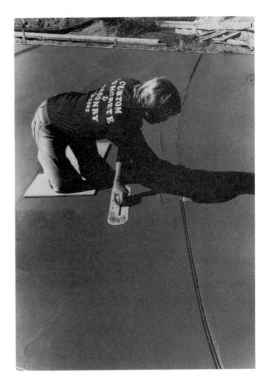

11-24 Both sides of this benderboard expansion joint have been edged from the finisher's location.

hand trowel. Work seamers back and forth along rough spots. You should not have to use a 2-×-4 guide, a seam's groove should be stable and supportive. If you experience trouble operating a walking seamer to smooth rough surfaces, take extension poles off and use it as a hand seamer. Simply finish seams as you did with edgers—out on the slab with knee boards.

BROOM FINISHING

The evolution of broom finishes was a concrete finisher's dream come true. Besides adding required traction to slabs, broom lines cover a lot of minor surface finish mistakes like slight trowel lines and some very light rough spots. Because broom bristles actually penetrate concrete surfaces, albeit not very deep, they are able to essentially roughen finishes to where small amounts of cream are able to fill in tiny holes and crevices (FIG. 11-25).

Walkways are generally broomed across their width for maximum traction capability. Patio slabs are broomed in line with their slope, generally in a direction perpendicular to houses or other structure walls they were poured against. Lines going in the direction of the slope on a patio slab, for instance, helps water rinse better because it flows toward the slab's lowest spot. If applied sideways, these lines would tend to dam water and possibly trap pieces of dirt and debris, making rinsing a bit more difficult and time-consuming.

11-25 A broom has been applied to the left side of this slab, resulting in a uniform finish of light penetrating lines.

The wetter concrete is, the deeper and more definitive broom lines will be. If your slab is a driveway on a very steep grade, for instance, you could apply a stiff-bristled broom to the concrete surface soon after fresno and edger work to effect an exceptionally rough finish (FIG. 11-26). This finish would greatly add traction for vehicles using a driveway in wet weather.

Concrete that has set up and is hard, might not accept much bristle penetration at all. In these cases, you might have to push and pull a soft bristled broom across the surface repeatedly to accomplish lines. In extreme situations, finishers have dunked broom bristles in buckets of water and then applied them wet.

Most slabs that have been finished on schedule will have enough moisture on their surface to allow neat, crisp, and sufficiently deep broom lines. Finishers can simply place their broom at the far end of a slab and then lightly pull it across the surface one time in a steady motion. Pushing a broom across slabs with accommodating surface texture (moisture) will leave lines that are far too deep. It will also amass cream build-up on the leading broom edge, causing blotches and wide scratches instead of thin clean lines.

In order to lift and move a broom head from outside formed areas to far reaching slab sides, you will have to attach extension poles to your broom handle. This is easily done by slipping an extension over the broom handle and then securing it with a wood screw inserted through an open hole on the extension and into the broom's wooden handle. If you do not have a soft-bristled push broom, regular concrete brooms are located at most tool rental facilities.

11-26 A broom was applied to this step while concrete was very wet, leaving behind deep heavy lines for extra traction when the surface is wet. An edger was used after brooming to smooth the perimeter and form a border around the center.

Pulling a broom across a concrete slab takes no special talent. All you do is raise the broom in the air, stretch it across the slab to its furthest point, gently lay it down and then pull it straight back toward you. If you see that bristle impressions are penetrating too deep, simply apply upward pressure on the poles to reduce the broom's weight on the slab. This will lessen its ability to dig into the surface and result in lighter lines (FIG. 11-27).

If you want all broom lines to run in the same direction, the most common pattern, you'll have to start at one end of the slab and work yourself across it, one broom length at a time. For a patio slab poured against the back of a house, for example, you would stand at a corner of a side form and back form edge while facing the house and lay down a broom line along the side edge; your broom will span from the side form in toward the slab. Once that pass is made, you simply lift and place the broom for another pass right next to the first, and so on until the entire slab is broomed.

If concrete is a little hard and one pass doesn't quite give the line depth desired, you can overlap each pass by half a broom length. Each pass will actually receive two broomings. Because of the span, your broom head operates at an angle as opposed to a flat plane in which brooms are normally operated for sweeping. This is of no consequence and will not cause any problems as long as it is pulled straight.

Other broom line designs or patterns can be applied. By wriggling a

11-27 To make straight lines, a broom must be pulled straight back. If lines appear to be penetrating too deep, simply apply upward pressure on the handle to reduce bristle weight on the surface. If lines are too light, apply downward pressure on the handle.

broom back and forth in a sideways motion, you can put on a wavy broom finish. As you pull the broom back with one hand, have the other positioned further up the handle closer to the broom head. Work the broom back and forth to create the pattern desired. Be sure to maintain the same sideways movement throughout each pass so that all passes will be symmetrical.

Another way to dress up broom finishes is to run an edger tool over all the edges after brooming. An edger will wipe out broom lines and leave a smooth border around a broomed slab. Before doing this, you must be certain that the smooth border will not hamper your ability to walk with good traction on slabs or walkways when they are wet.

In tight spots around slabs where a broom is impossible or exceptionally difficult to operate, simply remove the handle and use the broom head only (FIG. 11-28). In addition, you could just use a smaller foxtail broom.

A "windshield wiper" broom finish pattern is unique. It is applied with a fox tail broom that is moved over slabs in short, arm-length semi-circles about 2 to 3 feet in diameter. To do this, you'll have to carry a fox-tail broom with you as you are out on a slab with knee boards and trowels. After a section of concrete has been trowel finished, simply extend the broom out as far as your trowels were and, without moving your arm, bend your wrist from side to side while applying the broom. Pull the broom straight back while doing each section. The end result will

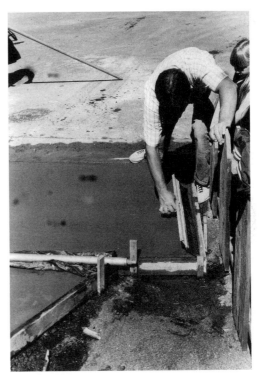

11-28 Broom applications in tight areas and close to form edges are not always easy with extension poles attached to brooms. For best access, remove poles and use brooms with their regular handle or without any handle at all, as is being done here to touch up a blemished area.

be a broomed slab consisting of semicircles all running in the same direction, with equal width and equal broom arc.

No matter what kind of broom finish design you apply, lines that are too deep can be troweled out and done over again later, when concrete is harder. If you notice this problem on your first broom pass, stop. Don't broom an entire slab that is too wet and then decide to trowel it all out and broom it again later. That is too much work. As soon as you notice that concrete is not quite ready for brooming, stop the process, trowel out what you have already done, and then wait until the surface has set up enough to accept the type of broom finish desired.

On the same note, if broom lines are too light, put downward pressure on the broom handle to help bristles penetrate deeper. If that doesn't work, pull and push the broom across each section a number of times using downward pressure as needed. As a last resort for slabs that are still too hard for ordinary brooming, dip broom bristles in a bucket of water and apply them soaking wet. Shake out excess water so the broom is not dripping continuous lines of water.

Chapter **12**

The last hour

*A*lthough there is still some work left to be done, concrete pouring and finishing should be complete by the time you reach this phase of the operation. Congratulations! Now, you can really sit down for a few minutes to catch your breath, have a cold drink, and admire your work. If you have followed directions up to this point, concrete should be completely finished and broomed, excess or spilled concrete cleaned up, and the tops of forms scraped off so only slight residual mud is left. Tools should be in good shape because they were rinsed with water after each use.

After your short rest, though, be prepared for some additional tasks. The last hour after concrete is trowel finished is filled with a number of small chores designed to protect new slabs and essentially complete jobs that leave work sites clean, tidy, and looking good.

PROTECTING CONCRETE

The longer concrete ages, the harder it gets. In fact, concrete takes about 99 years to reach its maximum strength and hardness. Rest assured, however, because concrete quickly reaches most of its strength (almost 99 percent) in only about two weeks. Along with short-term barrier protection to keep people and pets from walking on new slabs, you must consider how you will maintain moisture inside slabs for a couple of weeks. This time segment is about how long it takes for concrete to dry out and reach a hardened state where normal activities will not permanently scratch or mar surfaces. During this two-week "incubation" period, slabs are normally susceptible to cracks, scratches, and other damage, such as erosion from water falling off roof eaves.

Typically, cracks occur on fresh slabs because concrete dried out too fast. Mixtures need to cure rather slowly to allow all of the internal elements time to fully set up. This is evidenced by concrete's color changes. Fresh concrete is generally a dark color; professional finishers call it green. Cured concrete is white and is a definite sign that moisture has just about completely evaporated from green concrete. It takes about two weeks for new slabs to turn white and you should not walk on or use slabs until they have turned a uniform white color.

One way to help concrete cure slowly and maintain a sufficient moisture base is to keep it wet with water spray (FIG. 12-1). For two weeks, sprinkle slabs with a water spray at least six times a day. Ideally, they should be wet down every time any section starts to dry out; about once every two hours during the day. Concrete will cure under water, and you could even build a dam around slabs to maintain an inch or two of water on top of them. Although some special concrete jobs might need to be protected this way, building dams around slabs is quite impractical. Water spray from a garden hose is plenty good enough.

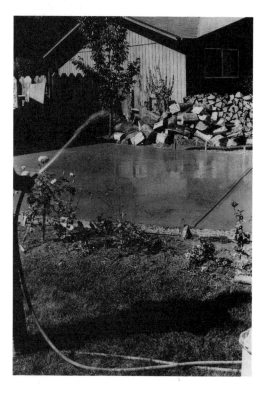

12-1 New concrete slabs should not be allowed to cure too fast, especially during warm summer weather.

Protecting slabs with water spray has a couple of minor drawbacks. One, of course, is the time required to actually do it. Someone must be available and has to remember to wet the slab down every two hours or so. The other drawback is most prevalent during warm summer weather. As the sun beats down on a slab, its surface temperature rises. Hitting it with a spray of cool water causes a thermo-shock, making the finish rapidly contract after expanding from the heat. This drastic change in surface temperature can cause "map cracks," a series of very fine, hairline surface cracks that wander all over finishes.

Map cracks are very shallow and do not pose strength-related problems, they are only cosmetic in nature. To avoid map cracks, make sure slabs are wet down consistently. Keep them wet and cool so that their surface temperature remains relatively stable.

Because professional concrete finishers cannot remain at job sites for two weeks to continually wet down slabs, they rely on special concrete cure and seal products to help slabs maintain required moisture (FIG. 12-2). This material is sprayed on finished slabs to seal surfaces so moisture does not rapidly evaporate. It is easily applied through a common variety garden sprayer. Concrete sealer products generally sell for around $12 a gallon with a gallon able to cover roughly 250 square feet.

12-2 A concrete sealer product helps slabs maintain sufficient moisture to ensure proper concrete curing. Here, spray is applied through a garden sprayer.

Concrete sealers are available at some lumberyards and hardware stores. Most often, though, sealers are acquired through ready mix concrete plants. The concrete truck driver might be able to bring it along to your job or you can go directly to the plant to buy it. Ask if you should bring along some plastic milk containers or other gallon jugs. Some concrete plants purchase this product in bulk supplies and will fill your containers from a 55-gallon drum. Because various sealer brands have different mixing or application ratios, you should consult the concrete plant dispatcher for specific application instructions.

Another way to help keep fresh slabs protected is with heavy-duty plastic tarps. Once concrete is finished and set up, lay a sheet of black plastic on top of it to help keep moisture from evaporating (FIG. 12-3). This may be one of the least expensive and care-free methods of allowing concrete to cure at its own rate. Be sure concrete is hard enough to accept a layer of plastic before putting it on, however. Wet concrete finishes can be scratched or gouged by wrinkles in plastic. Test finishes for hardness by touching them with your finger. If an imprint is made with your finger, a crease or wrinkle on plastic will do the same thing. If there is no noticeable impression, then plastic should not make any marks either. Secure plastic in place by laying 2×4s or bricks on top of it around the outside perimeter of the job, not on top of the concrete.

12-3 Plastic sheets protect new concrete from fall and winter rainfall and can be used in conjunction with concrete sealer during extra hot weather to prevent excessive moisture evaporation.

Plastic tarps are essential during winter—not so much to keep moisture in concrete, but to keep rain and cold away. Even semi-cured slabs can be ruined by raindrops and water runoff from roof eaves. Once concrete has sufficiently hardened, cover it with plastic stretched tight so wrinkles and creases are not allowed to rest on the finished surface. Plastic will lessen the impact of water drops and protect slabs against surface erosion.

Because concrete contains moisture, slabs must be protected against freezing temperatures. If moisture inside fresh concrete is allowed to freeze—essentially turning into ice—it will expand, causing cracks and might even result in chunks of the slab surface popping out. It would look like someone smacked different parts of the surface with a hammer and chipped out pockets of the finish. If possible, delay concrete pours until spring when weather has warmed to above freezing.

Professional concrete finishers protect slabs in freezing weather with plastic and straw. Once a slab has hardened enough to support a plastic cover, a thick layer of straw is spread over it to provide insulation. Plastic helps to keep concrete's own heat and moisture confined to the slab and a few inches of straw insulates against outside temperatures. In addition, a very slight amount of heat generated by straw's decomposition also helps to keep concrete from freezing.

CLEANUP

There is no getting around the fact that concrete work is messy. No matter how hard you try, mud will find its way to forms, tools, walls, preexisting slabs, and just about everything else located close to your job site. Cleanup should begin as soon as practical because fresh concrete is much easier to clean off than that which has sat around for a day or two.

Tools and equipment must have first cleanup priority. Unless they are cleaned promptly, concrete will set up on them making eventual cleaning a rigorous task or will simply render them useless for future jobs. Use a five-gallon bucket filled partially with water to clean finishing tools after each use. When the finishing job is complete, thoroughly clean tools with fresh water and a brush. Try to remove every spot of concrete, especially on blade faces and around bull float and fresno swivels.

Forms and stakes must also be cleaned. Water and a brush will make quick work of cleaning concrete off of them so that they are usable for future concrete jobs or other things, such as landscaping borders, patio covers, and the like. Be sure all nails are pulled out. To make this job easier, finishers fill a wheelbarrow full of water and put stakes in it. Once washed, they are neatly stacked and put away.

Areas around a job site need cleaning attention as well. The concrete truck driver had to clean out the chutes and you will have to remove whatever concrete residue resulted from that. For the most part, chutes are usually rinsed in the street, close to a curb. Use a square-point shovel to scoop up sand, aggregate, and cream, then wash away the rest with plenty of water. Although water pressure is generally enough to rinse away most concrete residue, you might need to use a stiff-bristled broom or brush to loosen stubborn accumulations.

Obstacles located in the middle of slabs also need to be cleaned before concrete has the opportunity to permanently set up on them. An old trowel or putty knife works good to scrape up loose concrete, while a wire brush does an excellent job of removing everything else. Leftover sand and crumbs of hardened concrete can be swept up with a whisk broom and dust pan.

Fresh concrete that is allowed to cure on top of old concrete will soon become a permanent fixture. Plan to clean preexisting concrete slabs as you would other obstacles. Use an old trowel or putty knife for the big stuff and then finish the job with a wire brush. You should do this right away. Waiting a day or two will allow new concrete to harden and make it very difficult to break loose and remove.

While using water to clean up areas in front of your house, be sure to control water streams so that they are not directed on top of new concrete. As accumulations of sand and aggregate are amassed, use a shovel to scoop them up into a wheelbarrow for proper disposal.

Heavy accumulations of excess concrete materials must be properly picked up and removed to an adequate disposal site. Use a wheelbarrow or five-gallon bucket, whichever is most practical, to transport sand, aggregate, and cement residue to a dumping site in your backyard or other suitable place (FIG. 12-4). If you have to take this debris to a landfill, put a piece of plastic down in the bed of your pickup truck or trailer before putting concrete residue in. This will help keep your truck or trailer bed clean.

12-4 Accumulations of concrete can be scooped up easily with a square-point shovel.

No matter where concrete truck chutes were rinsed or concrete overflow spilled, you can clean the area with water to make it look just like it did before the job was poured. All it takes is a little time and effort. The more neat and tidy job sites look, the better your concrete work will also look (FIG. 12-5).

PULLING FORMS

Forms and stakes are generally pulled the same day concrete is poured, provided concrete has set up to the point where it can support knee

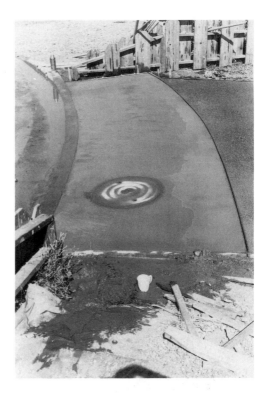

12-5 A few minutes of conscientious cleaning goes a long way toward a professional, crisp, and eye-appealing job well done.

boards without them making impressions on the surface. However, there are exceptions! Do not pull forms on the same day slabs were poured during cool, damp winter or autumn days and never pull forms on the same day exposed aggregate finishes were implemented (a custom concrete finish).

In these conditions, concrete has only had enough time to set up sufficiently on the surface. Underlying concrete might still be very wet and unstable. Pulling forms too early during cool, wet weather and on slabs exposed to a lot of water for exposed aggregate finishes, might allow unstable concrete to pull loose or slump down, ruining an otherwise perfect job. In these situations, plan to wait at least 2 or 3 days before pulling forms.

Pulling stakes and forms entails nothing more than extracting nails from stakes, loosening and removing them from the ground and then pulling forms horizontally away from slabs until they are free. Use a claw hammer or nail puller to take out all nails. Hit stakes on their sides with a small sledgehammer so they move back and forth along the sides of forms to loosen them. Pull stakes straight up to extract them from the ground. Stakes that have been driven a long way into the ground might require repeated hits with a hammer and some extra strong pulling to get them out.

Once all of the stakes have been removed, you can start to gently remove forms. Never pull forms straight up. Instead, always pull them horizontally away from slabs first. Often, forms are sort of attached to concrete by way of suction. Pulling straight up on forms can cause chunks of the concrete surface to come up with them. Although these problems can be easily repaired, it is really best to avoid them. By pulling forms away horizontally, you greatly reduce the chances of breaking loose chunks of surface concrete and forms will also come off much cleaner.

Start at one form end and gently break it loose from concrete by softly tapping on the board with a hammer horizontally. As you see this form start to separate from concrete, gently but firmly ease the form out and away. You might have to wriggle a form back and forth horizontally until the rest of its length breaks loose. Once it is free, pick it up and take it to the cleaning area. Be careful while carrying long forms that they are not brushed against the slab or knocked against anything that might break or be damaged. Clean forms with water and a brush.

REPAIRING EDGES

Sometimes, no matter how careful you are, a chunk of fresh concrete will come off of an edge as forms are pulled. This might be caused by a sliver of wood that has had concrete poured around it or from other unavoidable factors. Fortunately, this problem is easily corrected. When forms are pulled the same day as the pour, concrete is still green enough to allow edge repairs. All you'll need is a hand edger and a 2-×-4 block about a foot long, or long enough to extend past the damaged edge on both sides.

Quite simply, just put the 2-×-4 block in place next to the damaged edge so that it extends past it on both ends. It will sit just like the original form did. Then, scrape up the chunk of concrete that was pulled off or an amount of cream from the original form equal to what is missing. You might have to roll this wad of mud in your hand and add a few drops of water to it in order to get it into a creamy and usable consistency. Once that's done, put the wad of mud on top of the damaged edge and then operate your hand edger back and forth over it until it blends in and looks the same as the rest of the slab.

The wood block will support your effort so that you can maneuver a hand edger as you did with the original form in place. Remember to apply pressure against the 2-×-4 block with your free hand or knee because there will be no stakes there to support it.

If you pull forms a few days after a pour, edge repairs can still be done but will require a different technique. Because concrete will have hardened, you will need to coat the damaged area with concrete glue (Elmer's white glue will also work), and then gently replace the chunk that broke off. If the chunk just crumbled and is in no condition to be reattached, you might have to buy a sack of cement or concrete patch material and mix up enough material to make repairs. Be sure to follow label instructions on the package to ensure suitable repair results. Concrete patching products are sold at lumberyards, hardware stores, and through some ready-mix concrete companies.

OVERVIEW

Once all of your metal concrete tools have been washed and dried, you should consider coating them with some type of a rust-prevention material. Many finishers have had good results spraying their tools with a commercial lubricant, such as WD-40. A coating of this or a similar product prevents rust from forming on fresno and trowel blades and keep swivels and connectors in good working condition. Wood floats are sanded to remove rough textures and/or slivers.

Although professional concrete finishers take great pains to prepare and pour jobs perfectly, they sometimes have to contend with unpredictable cracks. Hog wire and/or rebar reinforcements help to prevent cracks from enlarging and separating, but you will need to apply any of a number of concrete patch products to seal cracks and prevent water from penetrating the voids created by them. Be sure to read product labels to ensure you purchase materials designed for the repairs you have in mind.

Index